5 STEPS TO A 5™

500 AP Physics B&C Questions to know by test day

Albert de Richemond

Craig C. Freudenrich

New York Chicago San Francisco Lisbon London Madrid Mexico City
Milan New Delhi San Juan Seoul Singapore Sydney Toronto

The ***McGraw-Hill*** *Companies*

1 2 3 4 5 6 7 8 9 10 FGR/FGR 1 9 8 7 6 5 4 3 2 1

ISBN 978-0-07-178072-8
MHID 0-07-178072-6

e-ISBN 978-0-07178073-5
e-MHID 0-07-178073-4

Library of Congress Control Number 2011936616

Series interior design by Jane Tenenbaum.

This book is printed on acid-free paper.

CONTENTS

ABOUT THE AUTHORS

Albert de Richemond is a professional engineer and teaches physics at Lehigh Carbon Community College in Schnecksville, PA. He also investigates medical device accidents and is an engineering consultant.

Craig C. Freudenrich holds a PhD and has 10 years of teaching experience in various sciences, including physics, biology, chemistry, anatomy, physiology, and astrobiology. He is currently a freelance science and education writer and serves as a science consultant in Durham, NC.

INTRODUCTION

Congratulations! You've taken a big step toward AP success by purchasing *5 Steps to a 5: 500 AP Physics B&C Questions to Know by Test Day*. We are here to help you take the next step and score high on your AP Exam so you can earn college credits and get into the college or university of your choice.

This book gives you 500 AP-style multiple-choice questions that cover all the-most essential course material. Each question has a detailed answer explanation. These questions will give you valuable independent practice to supplement your regular textbook and the groundwork you are already doing in your AP classroom. This and the other books in this series were written by expert AP teachers who know your exam inside out and can identify the crucial exam information as well as questions that are most likely to appear on the exam.

You might be the kind of student who takes several AP courses and needs to study extra questions a few weeks before the exam for a final review. Or you might be the kind of student who puts off preparing until the last weeks before the exam. No matter what your preparation style is, you will surely benefit from reviewing these 500 questions, which closely parallel the content, format, and degree of difficulty of the questions on the actual AP exam. These questions and their answer explanations are the ideal last-minute study tool for those final few weeks before the test.

Remember the old saying, "Practice makes perfect." If you practice with all the questions and answers in this book, we are certain you will build the skills and confidence needed to do great on the exam. Good luck!

—Editors of McGraw-Hill Education

CHAPTER 1

Vectors

1. Add the following vectors and give the resultant vector: $(5\hat{\mathbf{i}} + 9\hat{\mathbf{j}})$, $(6\hat{\mathbf{i}} - 3\hat{\mathbf{j}})$, $(2\hat{\mathbf{i}} - 12\hat{\mathbf{j}})$.

(A) 31
(B) 14, 3, −10
(C) $(13\hat{\mathbf{i}} - 6\hat{\mathbf{j}})$
(D) $(13\hat{\mathbf{i}} + 24\hat{\mathbf{j}})$
(E) $(14\hat{\mathbf{i}} + 3\hat{\mathbf{j}} - 10\text{ç})$

2. Why are North, South, East, and West not vectors?

(A) They are vectors.
(B) They state a direction.
(C) They state a direction but not a magnitude.
(D) They are scalars.
(E) They have no reference coordinates.

3. A temperature is a scalar because

(A) it is a vector.
(B) it measures heat.
(C) it only has one dimension and two directions.
(D) it states a magnitude but not a direction.
(E) it is a qualitative measure.

4. What is the difference between velocity and speed?

(A) There is no difference.
(B) Velocity is a vector, and speed is a scalar.
(C) Speed is a vector, and velocity is a scalar.
(D) Speed states only a magnitude, and velocity states a magnitude and direction.
(E) Velocity states only a magnitude, and speed states a magnitude and direction.

5. Why can a vector be expressed in an $R\,\theta\,\Phi$ system?

(A) It cannot be expressed in an $R\,\theta\,\Phi$ system.
(B) It cannot be expressed because it has no x, y, z directions.
(C) It can be expressed because a magnitude is expressed by R and a direction by θ and Φ.
(D) It can be expressed because $R\,\theta\,\Phi$ can be expressed as x, y, z components.
(E) It cannot be expressed because $R\,\theta\,\Phi$ is only used in special cases.

6. What is the magnitude of this vector, $100.\hat{\mathbf{i}} - 200.\hat{\mathbf{j}} + 1500.\mathbf{k}$?

(A) 1520
(B) 1516
(C) 1516.6
(D) 1800
(E) 1400

7. What is the angle made with the x axis by this vector, $1200.\hat{\mathbf{i}} - 200.\hat{\mathbf{j}}$?

(A) 9°
(B) 80°
(C) 9.46°
(D) 80.5°
(E) .167 radians

8. What is the dot product of two vectors, **A** and **B**?

(A) $|A|\cdot|B|$
(B) $|A|\cdot|B|\cdot\sin\theta$
(C) $|A|\cdot|B|\cdot\cos\theta$
(D) $|A|\cdot|B|\cdot\tan|\cdot\sin\theta$
(E) $|A|\cdot|B|\arctan|\cdot\sin\theta$

9. What is the cross product of two vectors, **A** and **B**?

(A) $|A|\cdot|B|\cdot\cos\theta$
(B) $|A|\cdot|B|\cdot\cos\theta$ directed perpendicular to the plane of **A** and **B**
(C) $|A|\cdot|B|\cdot\sin\theta$ directed perpendicular to the plane of **A** and **B**
(D) $|A|\cdot|B|\cdot\sin\theta$ directed perpendicular to the plane of **A** and **B** in the positive direction
(E) $|A|\cdot|B|\cdot\cos\theta$ directed perpendicular to the plane of **A** and **B** in the negative direction

10. Which of the following is a scalar?

(A) Force
(B) Acceleration
(C) Velocity
(D) Moles
(E) Angular velocity

11. What are the x and y components of a vector with a magnitude of 50, directed 33° from the x axis?

(A) 27.23 and 41.93
(B) 50 and 50
(C) 27 and 42
(D) 42 and 27
(E) 41.93 and 27.23

12. What are the x, y, and z components of a vector with a magnitude of 58, directed 33° from the x axis and 120° from the z axis?

(A) $49.\hat{\imath} + 32.\hat{\jmath} - 29.\mathbf{k}$
(B) $42.\hat{\imath} + 27.\hat{\jmath} - 29.\mathbf{k}$
(C) $49.\hat{\imath} + 27.\hat{\jmath} - 29.\mathbf{k}$
(D) $42.\hat{\imath} + 27.\hat{\jmath} - 29.\mathbf{k}$
(E) $42.\hat{\imath} - 27.\hat{\jmath} + 29.\mathbf{k}$

13. What is the resultant vector after adding $12.\hat{\imath} - 20.\hat{\jmath} + 150.\mathbf{k}$, $52.\hat{\imath} + 43.\hat{\jmath} - 1.\mathbf{k}$, and $-23.\hat{\imath} - 13.\hat{\jmath} - 61.\mathbf{k}$?

(A) $142.\hat{\imath} + 40.\hat{\jmath} + 136.\mathbf{k}$
(B) $41.\hat{\imath} + 10.\hat{\jmath} + 98.\mathbf{k}$
(C) $-41.\hat{\imath} - 10.\hat{\jmath} - 98.\mathbf{k}$
(D) 188
(E) 133

14. If a hiker heads north for 2.0 miles, northeast for 3.0 miles, west for 2.0 miles, and then south for 3.0 miles, what vector describes his final position?

(A) The hiker is lost
(B) The hiker is 1.1 miles from his starting point
(C) 0.1 mi $\hat{\imath}$ + 1.1 mi $\hat{\jmath}$
(D) 2.0 mi $\hat{\imath}$
(E) 1.0 mi $\hat{\jmath}$

15. Which of the following is a vector?

(A) Time
(B) Distance
(C) Force
(D) Speed
(E) Direction

16. Why are vectors needed?

(A) To define a location
(B) To define a force or moment
(C) To define acceleration
(D) (A), (B), and (C)
(E) Only (A)

17. What angle does the following vector, $53.0\hat{\imath} - 42.0\hat{\jmath} + 29.0\mathbf{k}$, make with the x axis?

(A) 90°
(B) 54.2°
(C) 35.8°
(D) 52°
(E) 38.4°

18. What angle does the following vector, $53.0\hat{\imath} - 42.0\hat{\jmath} + 29.0\mathbf{k}$, make with the z axis?

(A) 23.2°
(B) 66.8°
(C) 38.4°
(D) 21.5°
(E) 68.5°

19. What is the product of a vector and a scalar?

(A) A scalar
(B) A vector
(C) A magnitude
(D) The same vector
(E) A different direction

20. What is the product of two vectors?

(A) The angle between the two vectors
(B) A vector and a scalar
(C) A scalar
(D) A scalar or a vector
(E) A vector

21. What is the dot product of 13.0$\hat{\imath}$ and –42.0$\hat{\imath}$?

(A) 546.0
(B) –546.0
(C) 0.0
(D) 53.0$\hat{\imath}$
(E) –53.0$\hat{\imath}$

22. What is the cross product of 13.0$\hat{\imath}$ and –42.0$\hat{\imath}$?

(A) 546.0
(B) –546.0
(C) 0.0
(D) 53.0$\hat{\imath}$
(E) –53.0$\hat{\imath}$

23. What is the right-hand rule for vector cross products?

(A) It defines how screws should turn.
(B) It defines the vector coordinate system.
(C) It defines the direction of the cross product vector.
(D) It defines the positive or negative direction of the cross product vector.
(E) It defines the perpendicular direction of the cross product.

24. What is the dot product of 13.0$\hat{\imath}$ and –42.0$\hat{\jmath}$?

(A) 546.0
(B) –546.0
(C) 0.0
(D) 53.0$\hat{\imath}$
(E) –53.0$\hat{\imath}$

25. What is the cross product of $13.0\hat{\imath}$ and $-42.0\hat{\jmath}$?

(A) $-546.0\mathbf{k}$
(B) $546.0\mathbf{k}$
(C) 0.0
(D) 53.0
(E) −53.0

26. Air traffic controllers give "vectors" to inform pilots of the direction (degrees of the compass) in which they should fly. Is this a correct use of vectors? Why or why not?

(A) Yes, because the direction and the speed of the plane make up a vector
(B) No, because they need to know the height of the plane
(C) Yes, because the pilots know the height and speed of the plane
(D) No, because the only information provided is a direction without a magnitude
(E) No, because compass degrees are not scalars

27. Which of the following is a true mathematical operation?

(A) $A \times B = B \times A$
(B) $A + B = B + A$
(C) $A \times (B \cdot C)$
(D) $A \cdot (B \cdot C)$
(E) $A \times (B \times C) = (A \times B) \times C$

28. Why is $\mathbf{A} \cdot (\mathbf{A} \times \mathbf{B}) = 0$ a mathematical operation?

(A) This is not a correct mathematical operation.
(B) **A** and **B** must be 0.
(C) The cross product gives a vector perpendicular to **A** and $\sin 90° = 0$.
(D) The cross product gives a vector perpendicular to **A** and $\cos 90° = 0$.
(E) Acrtan $(\mathbf{A}/\mathbf{B}) = 0$.

29. For the following vectors, provide their *x*, *y*, *z* components, dot product, and cross product. Vector A has a magnitude of 20 N and is directed 40° from the *x* axis, 50° from the *y* axis, and 30° from the *z* axis. Vector B has a magnitude of 30 N and is directed 120° from the *x* axis, 30° from the *y* axis, and 300° from the *z* axis.

30. In the constellation Ursa Major, or the Big Dipper, one star, Alkaid, forms the end of the dipper handle and Merak forms the outside bottom corner of the dipper bowl. Alkaid is 138 light years away from the Earth, and Merak is 77 light years away from the Earth. When viewed from the Earth, the two stars are 25.6° apart. How far is Alkaid from Merak?

CHAPTER 2

Free-Body Diagrams and Equilibrium

31. Which of Newton's laws describes equilibrium?

(A) The first, which mathematically states that $\Sigma \mathbf{F} = 0$.
(B) The second, which mathematically states that $\Sigma \mathbf{F} = m\mathbf{a}$.
(C) The third, which mathematically states that $\mathbf{F} = -\mathbf{F}$.
(D) The fourth, which mathematically states that $\mathbf{F} = Gm_1m_2/r^2$.
(E) The fifth, which mathematically states that $dx/dt = v$.

32. A book is resting on a table. What forces are acting on the book?

(A) The weight of the book
(B) The weight of the book and the acceleration of gravity
(C) The supporting force of the table on the book
(D) The weight of the book and the pushback force of the table
(E) The weight of the book and the force of the table

33. When is the normal force not equal to the weight of an object?

(A) In space where there is no gravity
(B) When the object is being accelerated upward
(C) When the object is on an inclined plane
(D) When the object is supported by a string
(E) When it is moving along a flat surface

34. Why can a string hung from two opposite points never be made perfectly straight?

(A) A free-body diagram (FBD) of the string shows that the string tension must be infinite to balance the weight of the string.
(B) An FBD of the string shows that an angle must exist between the string and a perfectly straight line to balance the weight of the string.
(C) An FBD of the string shows that an angle must exist between the string and a perfectly straight line to allow vertical forces to balance the weight of the string.
(D) If the string were perfectly straight, it would break.
(E) Strings, ropes, and cable always hang in the shape of a helix.

35. A more general statement of Newton's first law is:

(A) $\Sigma \mathbf{F} = 0$
(B) $\Sigma \mathbf{F}_x = 0$, $\Sigma \mathbf{F}_y = 0$, and $\Sigma \mathbf{F}_z = 0$
(C) All the forces and moments must be zero
(D) All the forces added together must be zero
(E) $\Sigma \mathbf{F}_x = 0$, $\Sigma \mathbf{F}_y = 0$, $\Sigma \mathbf{F}_z = 0$, $\Sigma \mathbf{M}_x = 0$, $\Sigma \mathbf{M}_y = 0$, and $\Sigma \mathbf{M}_z = 0$

36. A torque is:

(A) A force applied at a distance
(B) A moment
(C) A force applied to a body that causes the body to rotate
(D) A pair of forces
(E) A force multiplied by a distance

37. Can a compressive force be carried by a rope?

(A) Only if the rope is very stiff
(B) No
(C) Yes
(D) If a rope is pushed, it tends to wriggle
(E) Generally, no

38. What is a friction force and how does it act?

(A) It is a force that acts along the plane.
(B) It is a force that opposes motion.
(C) It is a force that is proportional to the normal force that acts in opposition to the direction of motion.
(D) It is a force that is proportional to the normal force that acts in the direction of motion.
(E) It is a force that is proportional to a body's weight that acts in the direction of motion.

39. What is a Free Body Diagram (FBD)?

(A) An FBD shows the forces acting on a body.
(B) An FBD shows the forces and their locations on a body.
(C) An FBD shows the forces and accelerations and their locations acting on a body.
(D) An FBD shows the forces and torques and their locations acting on a body.
(E) An FBD shows the forces and velocities and their locations acting on a body.

40. What defines a lever arm?

(A) The distance parallel to the line of action of the force
(B) The distance parallel to the line of action of the force from the pivot point
(C) The distance perpendicular to the line of action of the force
(D) The distance perpendicular to the line of action of the force from the pivot point
(E) $\mathbf{F} = |F| \cos(\theta)x$

41. A pulley is supporting a 200-pound box. What is the tension in the rope?

(A) 100 pounds
(B) 200 pounds
(C) 50 pounds
(D) It depends on the angle that the rope makes with the pulley
(E) It depends on where the rope is placed on the box

42. A rope is strung between two cliffs. A mountain climber weighing 100 kg is halfway across the rope when the rope forms an angle of 10° with the horizontal. What is the tension in the rope?

(A) 980 N
(B) 498 N
(C) 980 kg
(D) 100 N
(E) 497 N

43. A 300-kg piano is being lifted into an upper-story window. Two pulleys are being used. At one point in the lift, one pulley rope on the left forms a 33° angle with the horizontal and forms a 63° angle with the horizontal on the right. What are the tensions in the two ropes?

(A) $T_{right} = 1{,}610$ N and $T_{left} = 2{,}633$ N
(B) $T_{right} = 1{,}610.0$ N and $T_{left} = 2{,}633.3$ N
(C) $T_{right} = 1{,}610.96$ N and $T_{left} = 2{,}633.92$ N
(D) $T_{right} = 1{,}610$ N and $T_{left} = 2{,}630$ N
(E) $T_{right} = 1{,}600$ N and $T_{left} = 2{,}600$ N

44. A 1-m pry bar is used to open a crate. It is forced 2 cm between the crate lid and the crate and is horizontal. If the pry bar tip is the fulcrum of the lever, and the end of the bar is pressed down with a 20-N force, what force is applied on the crate lid?

(A) 1 N
(B) 100 N
(C) 10 N
(D) 1,000 N
(E) 10,000 N

45. A 100-kg box is sitting on a 10° incline with a coefficient of friction of 0.5. How much must the incline be raised to make the box slide down?

(A) 30°
(B) 33°
(C) 26°
(D) 24°
(E) 15°

46. An astronaut floating in space uses an ordinary wrench to loosen a bolt on a massive satellite. What happens?

(A) He loosens the bolt.
(B) The satellite rotates around the bolt.
(C) The astronaut rotates around the bolt.
(D) Nothing happens.
(E) The astronaut moves in the direction of the force on the wrench.

47. A 5-kg baby carriage that is carrying a baby that weighs 5 kg rolls on all four wheels. The front wheels are 60 cm from the back wheels and 90 cm from the carriage handle. Assuming that the center of gravity for the baby and the carriage's center is equidistant between the front and back wheels, what force must the mother apply to tip the carriage back to lift the front wheels?

(A) 10 kg
(B) 100 N
(C) 5 kg
(D) 50 N
(E) 100 N·m

48. A six-strand pulley is used to raise a load weighing 1,000 kg. What is the tension on the pulley cables?

(A) 15,000 N
(B) 1,000 N
(C) 1,633 N
(D) 10,000 N
(E) 16,000 N

49. Newton published his laws of motion in what text?

(A) *Philosophiæ Naturalis Principia Physica*
(B) *Philosophiæ Naturalis Principia Mathematica*
(C) *Philosophiæ Naturalis Principia Il Movo*
(D) *Principia*
(E) *Principia Mathematica*

50. An object is supported by a spring attached to the ceiling. What are the forces acting on the object?

(A) The mass of the object in the downward direction and the supporting force of the spring in the upward direction
(B) The mass of the object in the downward direction and the supporting force of the spring in the downward direction
(C) The weight of the object in the upward direction and the supporting force of the spring in the downward direction
(D) The weight of the object in the downward direction and the supporting force of the spring in the upward direction
(E) The weight of the object and the supporting force of the spring

51. A box that is resting on an inclined plane has what forces acting on it?

(A) The weight of the box, the box's component forces perpendicular to and along the plane, and the friction force along the surface of the plane beneath the box
(B) The weight of the box, the box's component forces perpendicular to and along the plane, the resultant force of the plane on the box, and the friction force along the surface of the plane beneath the box
(C) The weight of the box and the friction force along the surface of the plane beneath the box
(D) The mass of the box and its acceleration along the plane
(E) The weight of the box, the resultant force of the plane on the box, and the friction force along the surface of the plane beneath the box

52. A crane is used to pick up a 50-m-long steel beam to place in a building. The beam is uniform, but the crane cable is placed 2 m off the center of the beam. How much vertical force must be placed on the guide rope to keep the beam level? The beam has a mass of 5 kg/m, and the guide rope is placed on the end of the shorter side of the beam.

(A) 213 N
(B) 213 kg
(C) 21 N
(D) 21 kg
(E) 106 N

53. A large chest is at the bottom of a lake. From its size, it has a buoyancy force of 200 N, but it is known to hold gold bullion with a mass of 200 kg. How much force is required to vertically pull the chest to the surface? How much force is required to pull it out of the water?

(A) 176 kg and 196 kg
(B) 176 N and 196 N
(C) 1,760 N and 1,960 N
(D) 1,960 N and 1,760 N
(E) 196 N and 1,760 N

54. A 2-m-long lever is placed under a 1,000-N rock. A fulcrum is placed at 0.5 m from the end of the rock, placing the lever at 30° above the horizontal. How much vertical force must be applied to the far end of the lever to lift the rock?

(A) 67 N
(B) 667 N
(C) 333 N
(D) 1,000 N
(E) 22 N

55. At what distance from the Earth is the center of mass for the Earth–Moon system? (The center of mass is the point between the bodies at which their masses balance. Assume that the Earth has a mass of 5.97×10^{24} kg, the Moon has a mass of 7.35×10^{22} kg, and they are 3.84×10^{8} m apart.)

(A) 5.00×10^{7} m from the Earth
(B) 1.92×10^{8} m from the Moon
(C) 4.73×10^{6} m from the Moon
(D) 4.73×10^{6} m from the Earth
(E) 1.92×10^{8} m from the Earth

56. A large 2,000-N cement block is pulled up a frictionless 43° incline using a dual-strand pulley that is attached to the block. One end of the rope is tied to the wall at the top of the plane while a man is pulling the other end of the rope. What is the tension in the pull of the rope if the rope is pulled parallel to the plane?

(A) 732 N
(B) 682 N
(C) 1,463 N
(D) 1,364 N
(E) 462 N

57. Three men are trying to position a heavy soda machine into a niche in a wall. Two men are pushing diagonally along the front of the machine. Where should the third man push so that the machine will move straight into the niche?

(A) At the top
(B) Pull from the rear
(C) Along a diagonal
(D) At the front center of the machine
(E) At either side

58. A helicopter is holding a 5,000-N bridge section 100 m over the spot where it is going to be placed, but the helicopter is slightly off. How much force must the spotter on the ground apply to the bridge section to move it 1 m into the proper position?

(A) 500 N
(B) 50 N
(C) 5 N
(D) 490 N
(E) 4.9 N

59. A large 2,000-N cement block is being pulled up a 43° incline with a single-strand pulley attached to the block. What is the tension in the pull rope if the rope is being pulled parallel to the plane? (Assume a coefficient of friction of 0.5 between the block and the plane.)

60. A crane has a 200,000-N counterweight, a 30-m boom, and a 12-strand pulley. The crane body is 4.0 m × 4.0 m, and the boom can be lowered to 5° from the horizontal. Neglecting the weight of the crane and boom, what weight can the crane lift without tipping?

CHAPTER 3

Kinematics

61. A 10-g penny is dropped from a building that is 125 m high. The penny is initially at rest. Approximately how long does it take the penny to hit the ground?

(A) 3.2 s
(B) 4.5 s
(C) 10 s
(D) 15 s
(E) 20 s

62. A car in a drag race started from rest and accelerated constantly to a velocity of 50 m/s when it reached the end of a 500-m road. What was the car's rate of acceleration?

(A) $-5.0\ m/s^2$
(B) $-2.5\ m/s^2$
(C) $0.5\ m/s^2$
(D) $2.5\ m/s^2$
(E) $5.0\ m/s^2$

63. An airplane is flying horizontally at a velocity of 50.0 m/s at an altitude of 125 m. It drops a package to observers on the ground below. Approximately how far will the package travel in the horizontal direction from the point that it was dropped?

(A) 100 m
(B) 159 m
(C) 250 m
(D) 1,020 m
(E) 1,590 m

64. A place kicker kicks a football at a velocity of 10.0 m/s from a tee on the ground at an angle of 30° from the horizontal. Approximately how long will the ball stay in the air?

(A) 0.0 s
(B) 0.6 s
(C) 0.8 s
(D) 1.0 s
(E) 1.8 s

65. This graph depicts the motion of an object. During which time interval is the object at rest?

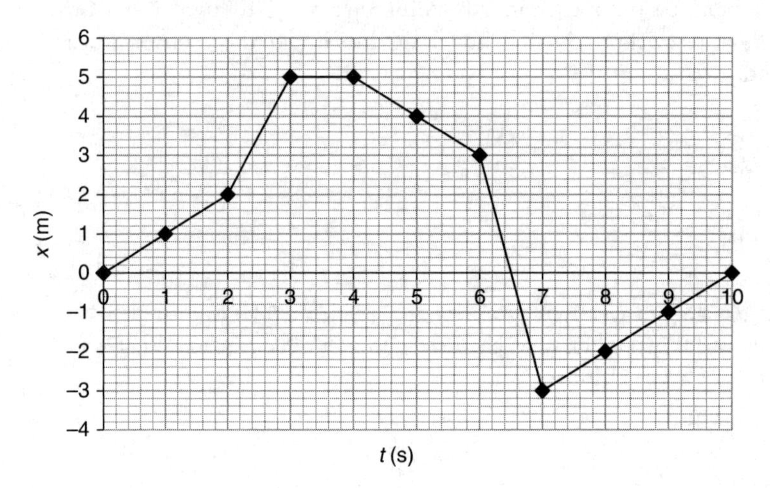

(A) 0–2 s
(B) 2–3 s
(C) 3–4 s
(D) 4–7 s
(E) 7–10 s

66. A car is traveling at an unknown velocity. It accelerates constantly over 5.0 s at a rate of 3.0 m/s^2 to reach a velocity of 30 m/s. What was the original velocity of the car?

(A) 1.0 m/s
(B) 5.0 m/s
(C) 10 m/s
(D) 15 m/s
(E) 20 m/s

67. A plane takes off from rest and accelerates constantly at a rate of 1.0 m/s for 5 min. How far does the plane travel in this time?

(A) 15 km
(B) 30 km
(C) 45 km
(D) 90 km
(E) 150 km

68. You drop a stone down a well and hear the echo 8.9 s later. If it takes 0.9 s for the echo to travel up the well, approximately how deep is the well?

(A) 40 m
(B) 320 m
(C) 405 m
(D) 640 m
(E) 810 m

69. If you throw a ball straight up into the air, which of the following expressions represents the time at which the ball reaches its maximum height?

(A) $t = v_f / v_0$
(B) $t = \Delta y / a$
(C) $t = a / (-v_0)$
(D) $t = a / \Delta y$
(E) $t = (-v_0) / a$

70. **[For Physics C Students Only]** The position of an object can be described by the function, $x(t) = 4t^2 + 6t + 2$. The units of t are in s and the units of x are in m. At 4 s, what is the direction and speed of the object?

(A) 90 m/s in the original direction of motion
(B) 38 m/s in the original direction of motion
(C) 0 m/s
(D) 38 m/s opposite the original direction of motion
(E) 90 m/s opposite the original direction of motion

71. On an airless planet, an astronaut drops a hammer from a height of 15 m. The hammer hits the ground in 1 s. What is the acceleration due to the gravity on this planet?

(A) 10 m/s^2
(B) 15 m/s^2
(C) 20 m/s^2
(D) 25 m/s^2
(E) 30 m/s^2

72. A car uniformly accelerates from rest at 3.0 m/s^2 down a 15-m track. What is the car's final velocity?

(A) 30 m/s
(B) 90 m/s
(C) 150 m/s
(D) 450 m/s
(E) 900 m/s

73. **[For Physics C Students Only]** An object's position with time is depicted in this graph. Based on that graph, at which time points will the object's velocity be zero?

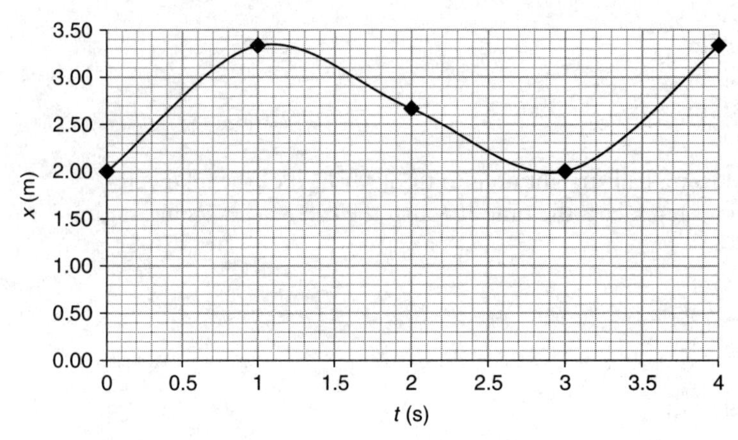

(A) 0 s and 2 s
(B) 0 s and 4 s
(C) 1 s and 2 s
(D) 1 s and 3 s
(E) 2 s and 4 s

74. At what time point does a projectile that is launched at an angle from the ground reach its maximum height?

(A) One-fifth of the total time in the air
(B) One-fourth of the total time in the air
(C) One-half of the total time in the air
(D) Three-fifths of the total time in the air
(E) Three-fourths of the total time in the air

75. A boy drops a stone from a cliff and counts th
the base. He counts 3 s. About how high is the

(A) 3 m
(B) 10 m
(C) 15 m
(D) 45 m
(E) 90 m

76. A boy is riding a bicycle at a velocity of 5.0 m/s. He applies
uniformly decelerates to a stop at a rate of 2.5 m/s^2. How lon
for the bicycle to stop?

(A) 0.5 s
(B) 1.0 s
(C) 1.5 s
(D) 2.0 s
(E) 2.5 s

77. A police officer finds 60 m of skid marks at the scene of a car crash. Assuming a uniform deceleration of 7.5 m/s^2 to a stop, at what was the initial velocity that the car was traveling when it started skidding?

(A) 20 m/s
(B) 30 m/s
(C) 45 m/s
(D) 60 m/s
(E) 90 m/s

he velocity–time graph of an object's motion is shown in this graph. At 0 s, what is the object's displacement relative to the initial time?

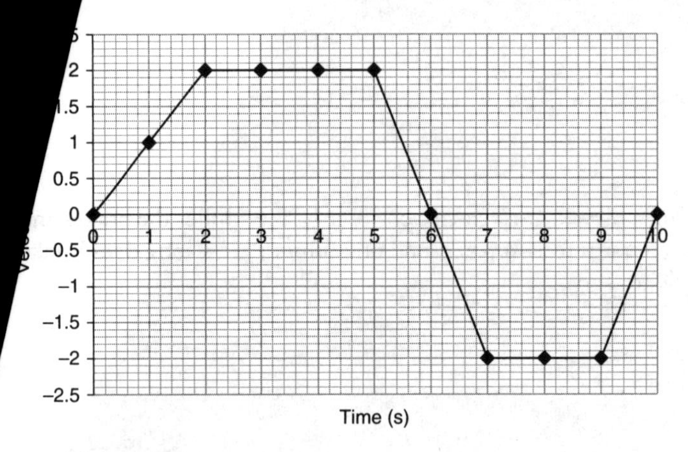

(A) 3 m
(B) 6 m
(C) 9 m
(D) −6 m
(E) −3 m

79. A projectile is launched with an unknown velocity at an angle of 30° from the horizontal of level ground. Which of the following statements is true?

(A) The horizontal component of velocity is less than the vertical component of velocity.
(B) The horizontal component of velocity is greater than the vertical component of velocity.
(C) Both the horizontal and vertical components of velocity are equal.
(D) The horizontal component of velocity is used to calculate the time that the projectile is in the air.
(E) The vertical component of velocity is used to calculate the range of the projectile.

80. From rest, a ball is dropped from the top of a building. It takes 4 s to hit the ground. Approximately how tall is the building?

(A) 40 m
(B) 90 m
(C) 100 m
(D) 160 m
(E) 200 m

81. The position–time graphs of five different objects are shown in these graphs. If the positive direction is forward, then which object is moving backward at a constant velocity?

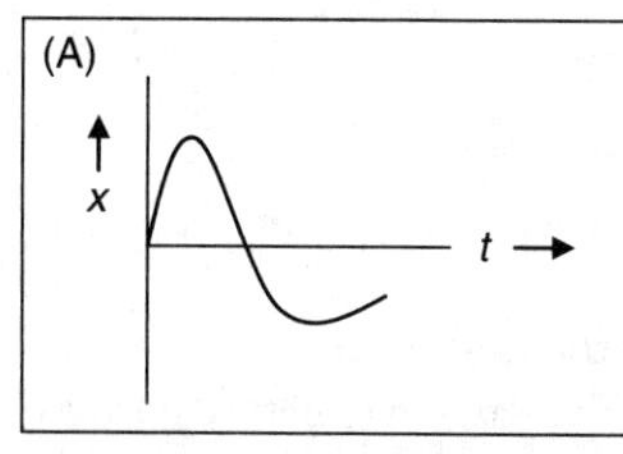

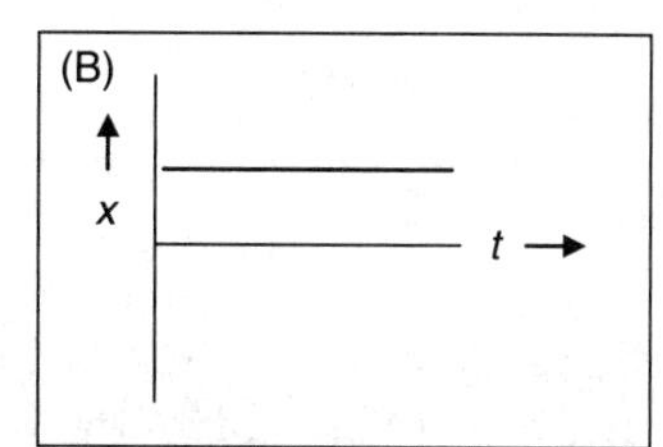

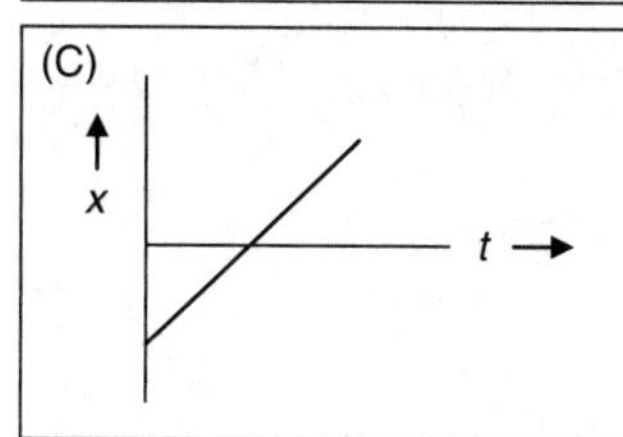

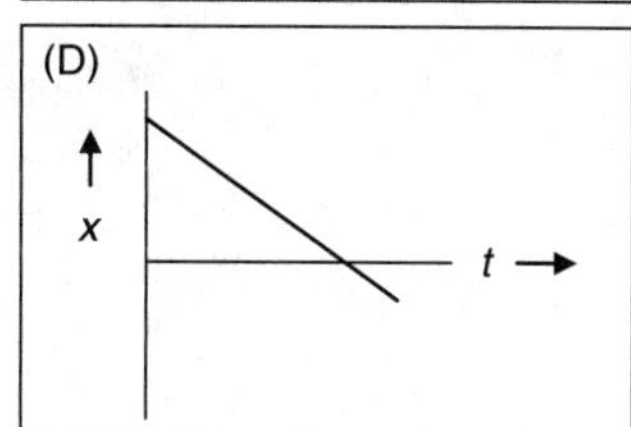

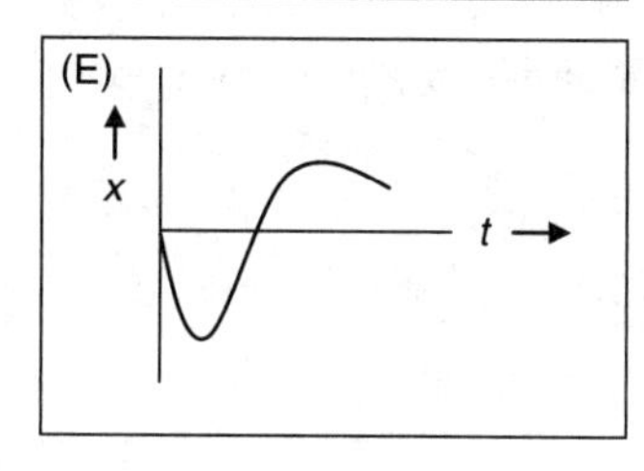

(A) Object A
(B) Object B
(C) Object C
(D) Object D
(E) Object E

82. A student launches projectiles with the same velocity, but at different angles (0–90°) relative to the ground. He measures the range of each projectile. Which angle pairs have the same range?

(A) 10° and 20°
(B) 30° and 45°
(C) 30° and 60°
(D) 45° and 60°
(E) 10° and 90°

83. Upon liftoff, a rocket accelerates at a constant 20 m/s^2 for 2 min. What is the rocket's velocity at that time?

(A) 40 m/s
(B) 140 m/s
(C) 1,200 m/s
(D) 2,400 m/s
(E) 4,800 m/s

84. A car is traveling at 30 m/s. The driver applies the brakes, and the car uniformly decelerates at 9 m/s^2. How far does the car travel before coming to a complete stop?

(A) 2 m
(B) 50 m
(C) 100 m
(D) 200 m
(E) 450 m

85. The position–time graph shown here is typical of which type of motion?

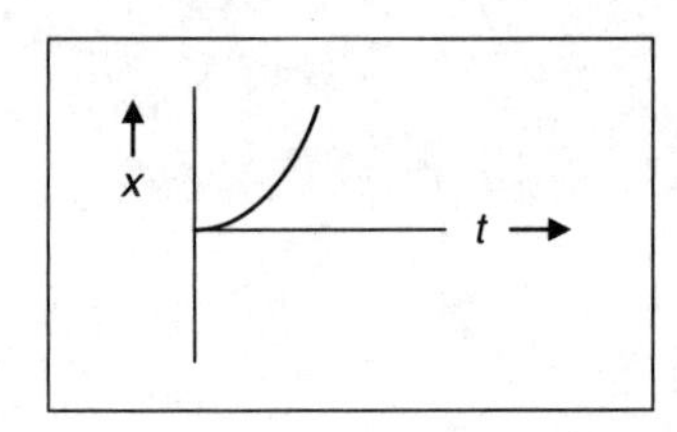

(A) Motion with a constant positive velocity
(B) Motion with zero velocity
(C) Motion with a constant positive acceleration
(D) Motion with zero acceleration
(E) Motion with a constant negative acceleration

86. If you drop a ball from a 100-m-tall building, approximately how far will the ball fall in 2 s?

(A) 10 m
(B) 20 m
(C) 40 m
(D) 50 m
(E) 100 m

87. An ice skater moving at 10 m/s comes to a complete stop in 0.5 s. What is the rate of acceleration?

(A) 5 m/s^2
(B) 10 m/s^2
(C) 20 m/s^2
(D) −20 m/s^2
(E) −10 m/s^2

88. The position–time graph shown here is typical of which type of motion?

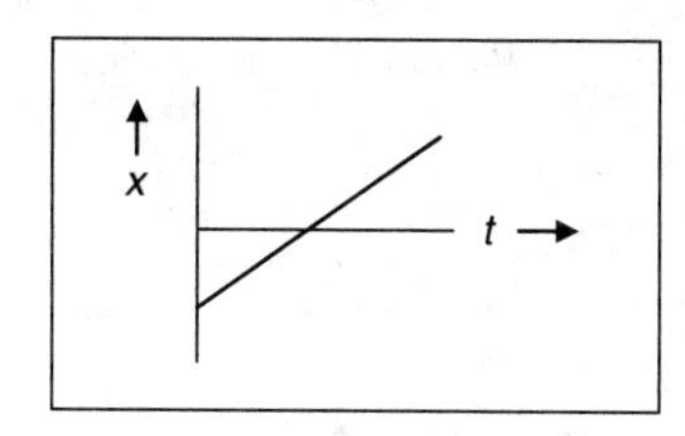

(A) Motion with a constant negative velocity
(B) Motion with zero velocity
(C) Motion with a constant positive acceleration
(D) Motion with zero acceleration
(E) Motion with a constant negative acceleration

89. An archer stands on a castle wall that is 45 m high. He shoots an arrow with a velocity of 10.0 m/s at an angle of 45° relative to the horizontal.

(a) Describe the path of the arrow.
(b) Determine the magnitude of the horizontal and vertical components of the arrow's velocity.
(c) Determine how much time it takes for the arrow to reach the ground.
(d) Determine the maximum range of the arrow.

90. The acceleration versus time of a bicycle rider is shown here.

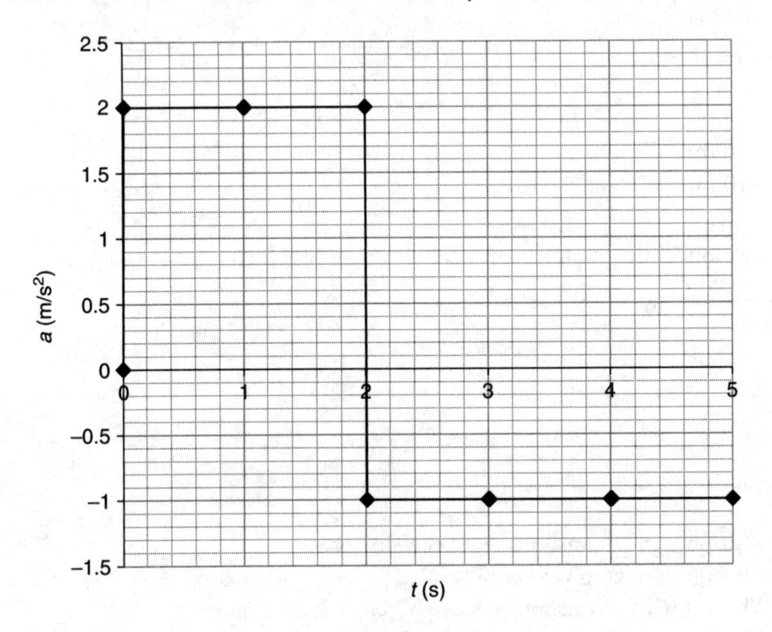

Assuming that the bicycle starts from rest ($x = 0$), do the following:

(a) Calculate the velocity–time data and plot a velocity–time graph.

(b) Calculate the position–time data and plot a position–time graph.

(c) Describe the motion of the bicycle. Assume that the positive direction is away from the starting point.

CHAPTER 4

Newton's Second Law

91. Two boys push a 5-kg box on a frictionless floor. James pushes the box with a 10.0-N force to the right. Louis pushes the box with an 8.0-N force to the left. What is the magnitude and direction of the box's acceleration?

(A) 0 m/s^2; the box does not move
(B) 0.4 m/s^2 to the right
(C) 0.4 m/s^2 to the left
(D) 1.6 m/s^2 to the right
(E) 1.6 m/s^2 to the left

92. A box of unknown mass (*m*) slides down a plane inclined at an angle (θ). The plane has a coefficient of friction (μ). Which of the following expressions would you use to calculate the rate of acceleration (*a*)?

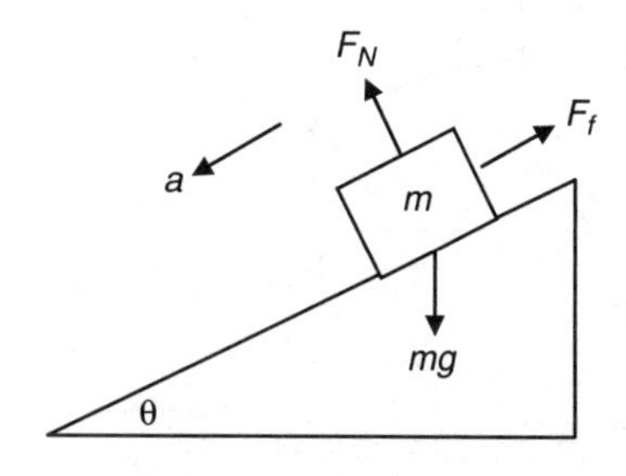

(A) $a = g(\sin\theta - \mu g \cos\theta)/m$
(B) $a = g(\cos\theta - \mu g \sin\theta)/m$
(C) $a = m(\sin\theta - \mu g \cos\theta)/g$
(D) $a = g(\sin\theta - \mu\cos\theta)$
(E) $a = g(\cos\theta - \mu\sin\theta)$

93. You pull down on one side of a rope stretched across a pulley. Attached to the other side of the rope is a 10-kg box. How much force must you pull down on the rope to get the box to accelerate upward at a rate of 10 m/s^2?

(A) 10 N
(B) 20 N
(C) 50 N
(D) 100 N
(E) 200 N

94. A jet takes off at an angle of 60° to the horizontal. The jet flies against a wind that exerts 1,000 N on it. The engines produce 20,000 N of thrust and the mass of the jet is 90,000 kg. What is the rate of the jet's acceleration in the horizontal direction?

(A) 0.1 m/s^2
(B) 0.2 m/s^2
(C) 1.0 m/s^2
(D) 5.0 m/s^2
(E) 10 m/s^2

95. This graph depicts the magnitudes and directions of two forces that act on an object. At which of the following times is the object accelerating?

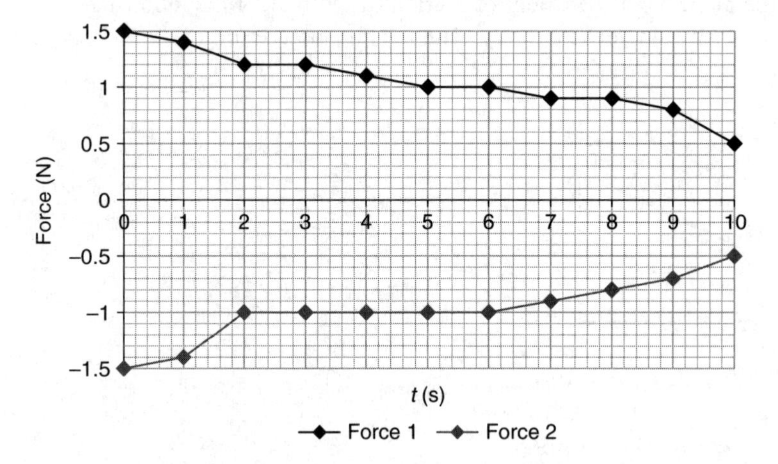

(A) 0 s and 1 s
(B) 2 s and 5 s
(C) 4 s and 8 s
(D) 1 s and 7 s
(E) 5 s and 10 s

96. **[For Physics C Students Only]** The motion of a 10-kg object can be described by the following equation, $x(t) = \frac{1}{3}t^3 - 2t^2 + 3t + 2$, where t is measured in s and x is measured in m. At what time will there be no net force acting on the object?

(A) 0 s
(B) 1 s
(C) 2 s
(D) 3 s
(E) 4 s

97. A boy pushes a 10-kg crate across the floor with a constant force of 10 N against a force of friction. The box accelerates at a rate of 0.1 m/s^2. What is the magnitude of the opposing frictional force?

(A) 0 N
(B) 1 N
(C) 5 N
(D) 9 N
(E) 10 N

98. A 10-kg crate is on a plane that is inclined at an angle of 45°. The coefficient of friction is 0.1, and the downward direction is positive. What is the approximate rate of the box's acceleration?

(A) 0 m/s^2
(B) 4.2 m/s^2
(C) 6.4 m/s^2
(D) 8.7 m/s^2
(E) 10 m/s^2

99. Two masses are connected by a rope across a pulley. The mass on the left of the pulley is 5 kg, while the mass on the right is 10 kg. If the positive direction of the pulley is counterclockwise, what is the magnitude and direction of the acceleration?

(A) 0 m/s^2
(B) −3.30 m/s^2
(C) −5.0 m/s^2
(D) 3.3 m/s^2
(E) 5.0 m/s^2

100. The acceleration of a 5-kg object over time is shown in this graph. What is the net force at 1 s?

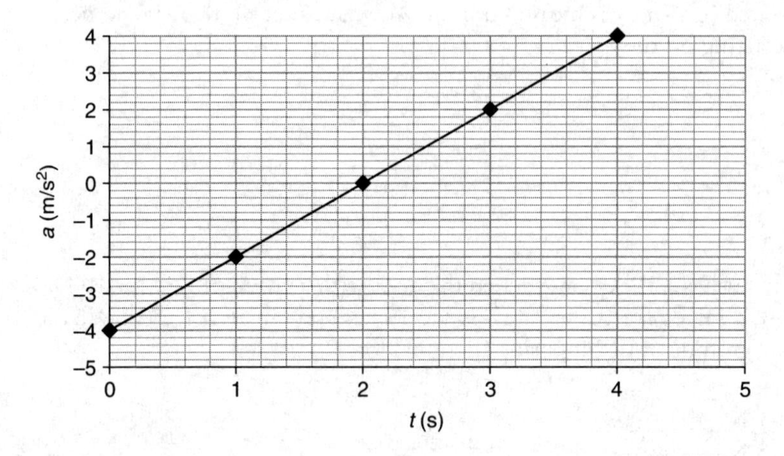

(A) −10 N
(B) −5 N
(C) −2.5 N
(D) 5 N
(E) 10 N

101. A skier weighing 70 kg pushes off from the top of a ski slope with a force of 105 N. The slope is inclined at 30°. Assuming that the slope is frictionless, what is the initial rate of the skier's acceleration?

(A) 0.5 m/s^2
(B) 1 m/s^2
(C) 2 m/s^2
(D) 7 m/s^2
(E) 11 m/s^2

102. A soldier fires a musket with a 1-m-long barrel. The gases from the exploding gunpowder exert a constant net force of 50 N on a 0.010-kg bullet as it travels through the musket barrel. What is the bullet's velocity as it leaves the musket barrel?

(A) 10 m/s
(B) 100 m/s
(C) 1,000 m/s
(D) 10,000 m/s
(E) 100,000 m/s

103. **[For Physics C Students Only]** An object's position with time is depicted in this graph. At which time will there be no net force acting on the object?

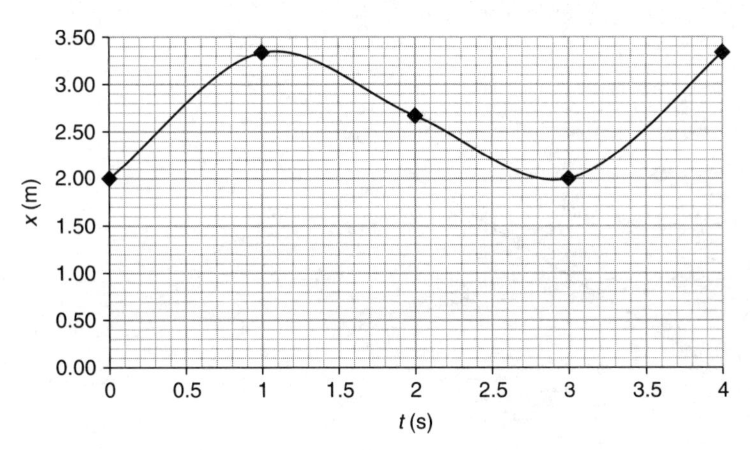

(A) 0 s
(B) 1 s
(C) 2 s
(D) 3 s
(E) 4 s

104. A block with a mass of 30 kg is located on a frictionless tabletop. This block is connected by a rope to another block with a mass of 10 kg. The rope is looped through a pulley on the table's edge so that the less massive block is hanging over the edge. Consider counterclockwise rotation of the pulley as positive. What is the rate of acceleration of the larger block across the table?

(A) -0.5 m/s^2
(B) -1 m/s^2
(C) -2.5 m/s^2
(D) 2.5 m/s^2
(E) 1 m/s^2

105. A jet flies at level flight. The engines produce a total of 20,000 N of thrust, the jet's mass is 50,000 kg, and it accelerates at 0.3 m/s^2. What is the magnitude of the air resistance against which the jet flies?

(A) 1,000 N
(B) 3,000 N
(C) 5,000 N
(D) 10,000 N
(E) 15,000 N

106. A boy is pushing a 50-kg crate across a frictionless surface. The velocity is changing with time as shown in this graph. What is the magnitude of the force that the boy applies to the crate?

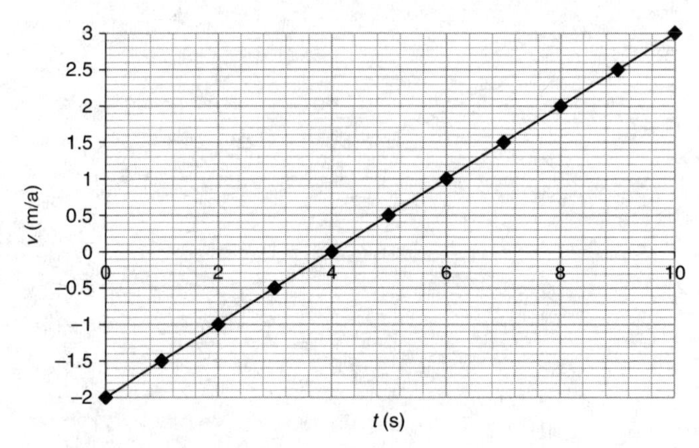

(A) 5 N
(B) 10 N
(C) 15 N
(D) 20 N
(E) 25 N

107. A car with a mass of 1,000 kg travels at 30 m/s. The driver applies his brakes for a uniform deceleration and comes to a complete stop in 60 m. Assuming that the forward motion is positive, what is the magnitude and direction of the net force acting on the car?

(A) 7,500 N
(B) 5,000 N
(C) −1,000 N
(D) −5,000 N
(E) −7,500 N

108. A bowler applies a constant net force of 100 N on a 5-kg bowling ball over a time period of 1.5 s before he releases the ball. The ball starts from rest. What is its final velocity?

(A) 5 m/s
(B) 10 m/s
(C) 20 m/s
(D) 30 m/s
(E) 40 m/s

109. A place kicker kicks a football with an unknown velocity at an angle of 45° from the horizontal of level ground. The football travels against the wind. Which of the following statements is true?

(A) The time that it takes the football to hit the ground will be increased.
(B) The horizontal velocity will be less than it would have been had there been no wind.
(C) The time that it takes the football to hit the ground will be decreased.
(D) The football will travel further than it would have had there been no wind.
(E) The height of the football will be less than it would have been had there been no wind.

110. This graph depicts the velocity of a skydiver over time during a free fall. Which of the following statements is true?

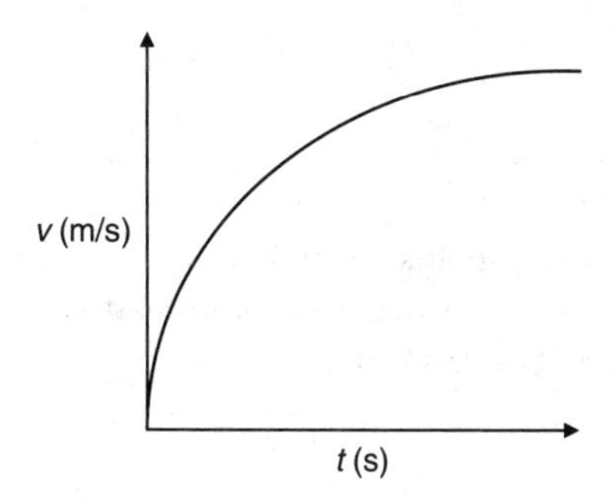

(A) The net forces acting on the skydiver increase until the acceleration reaches its maximum and the velocity becomes constant.
(B) The net forces acting on the skydiver decrease until the acceleration reaches zero and the velocity becomes constant.
(C) There are no net forces acting on the skydiver.
(D) The net forces acting on the skydiver increase until the acceleration reaches zero and the velocity becomes constant.
(E) The net forces acting on the skydiver decrease until the acceleration reaches zero and the velocity becomes zero.

111. A girl pushes a 10-kg box from rest across a floor with a force of 50 N. The force of friction opposing her is 45 N. If the box uniformly accelerates to a final velocity of 2.0 m/s^2, how long did it take to get to that velocity?

(A) 1 s
(B) 2 s
(C) 3 s
(D) 4 s
(E) 5 s

112. A rocket goes from rest to 9.6 km/s in 8 min. The rocket's mass is 8.0×10^6 kg. Assuming a constant acceleration, what is the net force acting on the rocket?

(A) 1.6×10^5 N
(B) 9.6×10^5 N
(C) 9.6×10^6 N
(D) 1.0×10^7 N
(E) 1.6×10^8 N

113. A car travels at a constant velocity of 30 m/s. The mass of the car is 1,000 kg. The car's engine produces 1,000 N of thrust against the force of friction of the road. What is the coefficient of friction of the road?

(A) 0.1
(B) 0.2
(C) 0.3
(D) 0.4
(E) 0.5

114. A 1,000-kg car is traveling at 30 m/s. The driver applies the brakes, which exert a constant 9,000 N of force. The car uniformly decelerates to a complete stop. How far does the car travel before coming to a complete stop?

(A) 2 m
(B) 50 m
(C) 100 m
(D) 200 m
(E) 450 m

115. This position–time graph is typical of which type of motion?

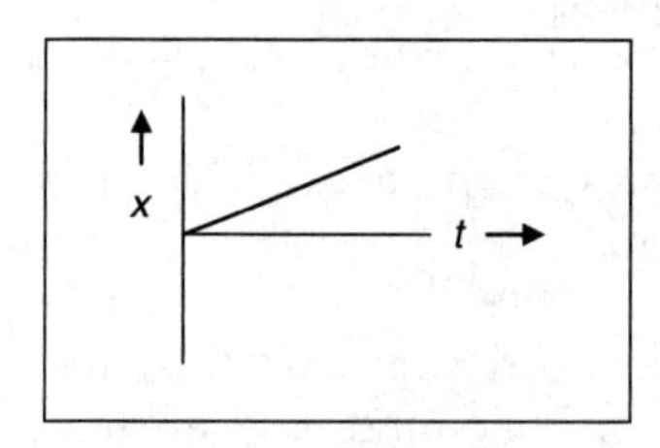

(A) Motion of an object with an increasing net force acting upon it
(B) Motion with zero velocity
(C) Motion with a constant positive net force acting upon it
(D) Motion of an object with no net force acting upon it
(E) Motion of an object with negative net force acting upon it

116. A 70-kg runner accelerates uniformly from rest to 10 m/s in 0.5 s. What is the net force acting upon him?

(A) −7,000 N
(B) −1,400 N
(C) 0 N
(D) 1,400 N
(E) 7,000 N

117. A 50-kg ice skater comes to a complete stop by applying a force of friction of 1,000 N opposite her direction of motion. The deceleration is uniform over a period of 0.5 s. What was her initial velocity?

(A) 5 m/s
(B) 10 m/s
(C) 20 m/s
(D) −20 m/s
(E) −10 m/s

118. The acceleration–time graph of an object's motion is shown in this figure. At what time will the forces acting on it be balanced?

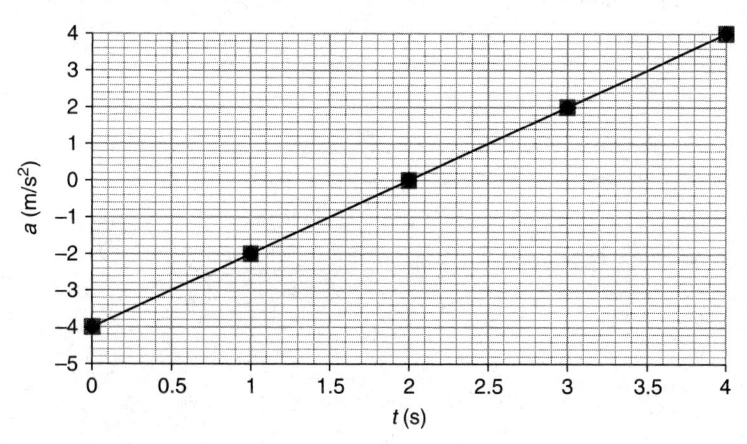

(A) 0 s
(B) 1 s
(C) 2 s
(D) 3 s
(E) 4 s

119. An archer stands on a castle wall that is 45 m high. He shoots an arrow with a velocity of 10.0 m/s at an angle of 45° relative to the horizontal. The arrow is shot into a constant headwind that exerts a force of 0.05 N on the arrow. The arrow has a mass of 0.050 kg.

(a) Determine the magnitude of the horizontal and vertical components of the arrow's initial velocity.
(b) Determine how much time it takes for the arrow to reach the ground.
(c) Determine the maximum range of the arrow into the wind.
(d) How much farther would the arrow go if the wind wasn't blowing?

120. A 10-kg box slides down a plane inclined at an angle ($\theta = 30°$). The plane has a coefficient of friction ($\mu = 0.1$). The box starts from rest and slides down the plane for 2.0 s.

(a) Draw a free-body diagram of this situation and label all the forces on the box.
(b) Calculate the force of friction on the box.
(c) Calculate the acceleration of the box.
(d) Calculate the final velocity of the box.
(e) Calculate the distance that the box moves down the plane in the given time interval.

CHAPTER 5

Momentum

121. A 70-kg man runs at a constant velocity of 2 m/s. What is his momentum?

(A) 35 kg m/s
(B) 68 kg m/s
(C) 70 kg m/s
(D) 72 kg m/s
(E) 140 kg m/s

122. A 10-N force is applied to a hockey puck over a period of 0.1 s. What is the change in momentum of the hockey puck?

(A) 1.0 kg m/s
(B) 9.9 kg m/s
(C) 10 kg m/s
(D) 10.1 kg m/s
(E) 100 kg m/s

123. A 1,000-kg car moving at a constant velocity of 11.0 m/s strikes a concrete barrier and comes to a complete stop in 2.0 s. What is the force acting on the car?

(A) −5,500 N
(B) −180 N
(C) 0.02 N
(D) 180 N
(E) 5,500 N

124. A 1.0-kg ball moving at 10 m/s strikes a wall and bounces back. The collision is perfectly elastic. What is the ball's momentum after the collision? (Assume a frictionless surface.)

(A) −100 kg m/s
(B) −10 kg m/s
(C) −1.0 kg m/s
(D) 10 kg m/s
(E) 100 kg m/s

125. A 10-kg box is sliding across an ice rink at 10 m/s. A skater exerts a constant force of 10 N against it. How long will it take for the box to come to a complete stop?

(A) 0.5 s
(B) 1.0 s
(C) 10 s
(D) 50 s
(E) 100 s

126. Two balls of equal mass collide in a perfectly elastic collision. Ball A moves to the right at 10 m/s. Ball B moves to the left at 5 m/s. After the collision, Ball B moves to the right at 3 m/s. What is the velocity of Ball A after the collision? (Assume a frictionless surface.)

(A) −10 m/s
(B) −8.0 m/s
(C) −5.0 m/s
(D) 2 m/s
(E) 3 m/s

127. A 140-kg fullback is running with the football at 10 m/s. A 70-kg defender runs at him in the opposite direction at 5 m/s. The defender wraps his arms around the fullback. What is the velocity of the two players after the collision? (Assume a frictionless surface.)

(A) −10 m/s
(B) −5 m/s
(C) 0 m/s
(D) 5 m/s
(E) 10 m/s

128. There are two billiard balls each with a mass of 200 g. A pool player shoots Ball A with a velocity of 1.0 m/s at Ball B, which is at rest. After the collision, Ball A stops and Ball B travels off in a straight line. What is the velocity of Ball B after the collision? (Assume a frictionless surface.)

(A) −1.0 m/s
(B) −0.2 m/s
(C) 0 m/s
(D) 0.2 m/s
(E) 1.0 m/s

129. A bullet has a mass of 50 g and its momentum is 25 kg m/s. What is the velocity of the bullet?

(A) 2×10^{-3} m/s
(B) 5×10^{-1} m/s
(C) 2.5×10^{1} m/s
(D) 5×10^{1} m/s
(E) 5×10^{2} m/s

130. **[For Physics C Students Only]** Two objects are moving around a fixed reference point. Object 1 is located at 2.0 m (x_1) and has a mass (m_1) of 10 kg. Object 2 is located at 6.0 m (x_2) and has a mass (m_2) of 20 kg. Where is the center of mass of the system?

(A) 2.1 m
(B) 4.0 m
(C) 4.7 m
(D) 5.0 m
(E) 5.9 m

131. A 150-kg halfback is running down the field carrying the ball at a velocity of 5 m/s. A 50-kg receiver from the opposing team is running at him in the opposite direction. The receiver hopes to tackle and stop him in an inelastic collision. Will the receiver be able to stop him? And, if so, at what velocity must he run?

(A) Yes, −15 m/s
(B) Yes, −10 m/s
(C) Yes, −5 m/s
(D) Yes, 0 m/s
(E) No, the receiver cannot stop the fullback

132. A batter applies a constant force of 10.0 N over a period of 5 m/s when he strikes a baseball. The mass of the baseball is 145 g. What is the change in velocity of the baseball?

(A) 0.100 m/s
(B) 0.345 m/s
(C) 0.500 m/s
(D) 2.00 m/s
(E) 2.90 m/s

133. What is the momentum of an electron moving at 90 percent of the speed of light?

(A) 3.3×10^{-31} kg·m/s
(B) 2.5×10^{-31} kg·m/s
(C) 3.3×10^{-22} kg·m/s
(D) 2.5×10^{-22} kg·m/s
(E) 3.3×10^{-20} kg·m/s

134. A 260-g cue ball moving at 1.0 m/s strikes a 150-g numbered ball at rest. The collision is elastic and the cue ball stops. What is the velocity of the numbered ball?

(A) 0.6 m/s
(B) 1.2 m/s
(C) 1.5 m/s
(D) 1.7 m/s
(E) 2.0 m/s

135. A 750-kg aircraft is flying level at 100 m/s. A tailwind blows for 2 min and the aircraft's speed increases to 120 m/s. What was the force of the tailwind?

(A) 125 N
(B) 250 N
(C) 2,500 N
(D) 5,000 N
(E) 7,500 N

136. **[For Physics C Students Only]** This graph depicts the motion of a box being pushed across the floor. Which of the following statements describes the force upon the box?

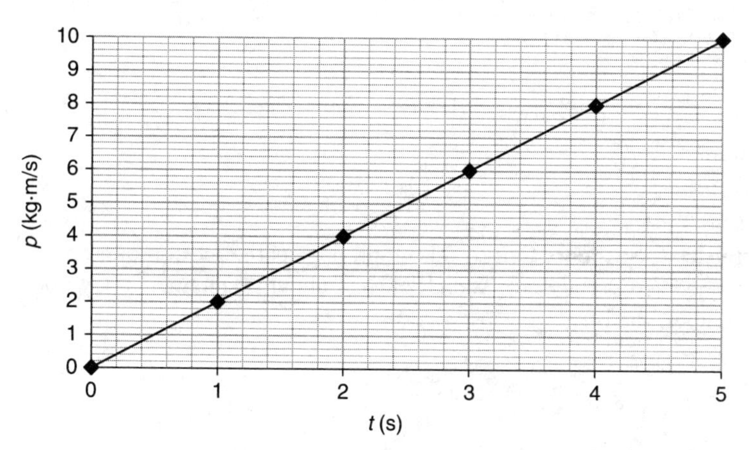

(A) The force on the box is a constant 0.5 N.
(B) The force on the box is a constant 1 N.
(C) The force on the box is a constant 2 N.
(D) The force on the box is 1 N and increasing.
(E) The force on the box is 2 N and increasing.

137. A force of 2.0 N exerted on an object for 100 m/s increases the object's velocity by 1.0 m/s. What was the mass of the object?

(A) 0.1 kg
(B) 0.2 kg
(C) 1.0 kg
(D) 20 kg
(E) 200 kg

138. A 70-kg stuntman free falls from a building for 5 s and hits an air bag. The air bag exerts a force on him over a time period of 2.5 s and he comes to a complete stop. What was the force exerted by the air bag?

(A) 100 N
(B) 275 N
(C) 475 N
(D) 875 N
(E) 1,000 N

139. Two railroad cars (2×10^4 kg each) are traveling in the same direction along a railroad track. Car A is traveling at 14 m/s, and Car B is traveling at 10 m/s. The two cars collide. What will the velocity of the combined cars be after the collision?

(A) 10 m/s
(B) 12 m/s
(C) 14 m/s
(D) 20 m/s
(E) 24 m/s

140. **[For Physics C Students Only]** This graph depicts the motion of a box being pushed across the floor. Which of the following statements describes the force upon the box?

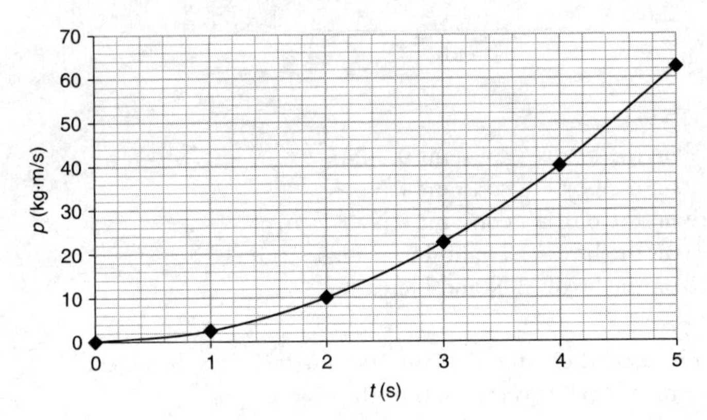

(A) The force on the box is constant.
(B) The force on the box is decreasing at a constant rate.
(C) The force on the box is increasing at a constant rate.
(D) The acceleration of the box is constant.
(E) The force on the box is zero.

141. Two skaters, a 70-kg man and a 50-kg woman, are skating on an ice rink. They "push off" from each other and move in opposite directions. After the push off, the woman moves with a velocity of 2.5 m/s. What is the velocity of the man?

(A) −2.5 m/s
(B) −1.8 m/s
(C) −1.3 m/s
(D) 0 m/s
(E) 2.5 m/s

142. To have a momentum of 10 kg·m/s, at what velocity must a 0.5-kg object be traveling?

(A) 1.0 m/s
(B) 5.0 m/s
(C) 10 m/s
(D) 20 m/s
(E) 50 m/s

143. A cue ball with a mass of 250 g travels at 1.0 m/s and hits a numbered ball with a mass of 170 g at rest. The numbered ball moves off at an angle of 45°, while the cue ball moves off at an angle of −45°. At what velocities do the balls move?

(A) The cue ball and the numbered ball move at 1.0 m/s.
(B) The cue ball moves at 1.0 m/s, and the numbered ball moves at −1.0 m/s.
(C) The cue ball moves at −0.71 m/s, and the numbered ball moves at 1.04 m/s.
(D) The cue ball moves at 1.04 m/s, and the numbered ball moves at −1.04 m/s.
(E) The cue ball moves at 0.71 m/s, and the numbered ball moves at 1.04 m/s.

144. A 2,000-kg airplane flies at 343 m/s. What is its momentum?

(A) 0.17 kg·m/s
(B) 5.8 kg·m/s
(C) 6,900 kg·m/s
(D) 6.9×10^4 kg·m/s
(E) 6.9×10^5 kg·m/s

145. A marksman fires a 7.5-g bullet from a 1.2-kg handgun. The bullet travels away at 365 m/s. At what velocity does the handgun recoil?

(A) −2.4 m/s
(B) −1.2 m/s
(C) 0 m/s
(D) 1.2 m/s
(E) 2.4 m/s

146. Two air cars weighing 500 g each are on a frictionless track. Car A is moving at 0.2 m/s when it collides into Car B at rest. The two cars get stuck together. At what velocity will the combined cars move?

(A) −0.1 m/s
(B) −0.05 m/s
(C) 0 m/s
(D) 0.05 m/s
(E) 0.1 m/s

147. A rifle fires a 4.0-g bullet at a velocity of 950 m/s. If the bullet is in the rifle barrel for only 0.1 s, what force does the rifle exert on the bullet?

(A) 18 N
(B) 28 N
(C) 38 N
(D) 48 N
(E) 58 N

148. A 1,000-kg cannon fires a 15-kg cannon ball. The cannon is mounted on a carriage that allows it to recoil at −1.5 m/s. What is the velocity of the cannon ball?

(A) 10 m/s
(B) 50 m/s
(C) 100 m/s
(D) 200 m/s
(E) 500 m/s

149. A cue ball with a mass of 250 g travels at 1.0 m/s and hits a numbered ball with a mass of 170 g at rest. The numbered ball moves off at an angle of 30°, while the cue ball moves off at an angle of −60°. At what velocities do the balls move? (Show your work.)

150. An 800-kg car is traveling along a wet road at a velocity of 25.5 m/s. A 1,000-kg car is traveling along the same road in the same direction at 34.7 m/s. The two cars collide and lock together. Answer the following questions. (Show your work.)

(a) The two interlocked cars proceed at what velocity after the collision?
(b) If the coefficient of -kinetic friction between the tires of the cars and the wet pavement is 0.7, what is the force of friction?
(c) How long does it take for the two interlocked cars to come to a complete stop on the wet pavement?

CHAPTER 6

Energy Conservation

151. A 70-kg man runs at a constant velocity of 2 m/s. What is his kinetic energy?

(A) 35 J
(B) 70 J
(C) 105 J
(D) 140 J
(E) 210 J

152. A 10-N force is applied horizontally on a box to move it 10 m across a frictionless surface. How much work was done to move the box?

(A) 0 J
(B) 10 J
(C) 50 J
(D) 100 J
(E) 1,000 J

153. A constant net force of 500 N moves a 1,000-kg car 100 m. If the car was initially at rest, then what is the car's final velocity?

(A) 1 m/s
(B) 5 m/s
(C) 10 m/s
(D) 20 m/s
(E) 100 m/s

154. A mover uses a pulley system to lift a grand piano with a mass of 500 kg from the ground to a height of 10 m. What is the change in the piano's gravitational potential energy?

(A) 5×10^2 J
(B) 5×10^3 J
(C) 5×10^4 J
(D) 5×10^5 J
(E) 5×10^6 J

155. A boy pulls a 10-kg box across an ice rink across a distance of 50 m. He exerts a constant force of 10 N on a rope attached to the box at an angle of 60°. How much work has he done on the box?

(A) 50 J
(B) 100 J
(C) 200 J
(D) 250 J
(E) 500 J

156. A 5-kg ball rolls down a hill from a height of 10 m. The ball starts from rest. What is the ball's velocity at the bottom of the hill?

(A) 5 m/s
(B) 10 m/s
(C) 14 m/s
(D) 24 m/s
(E) 30 m/s

157. An archer pulls a bowstring back a distance of 1.5 m with 450 N of force. The arrow has a mass of 20 g. When he releases the string, what is the velocity of the arrow when it leaves the bow?

(A) 100 m/s
(B) 200 m/s
(C) 260 m/s
(D) 360 m/s
(E) 500 m/s

158. A car traveling at 30 m/s must stop within 10 m to avoid a crash. What magnitude of force must the brakes exert on the car?

(A) 1.5×10^3 N
(B) 3.0×10^3 N
(C) 4.5×10^3 N
(D) 1.5×10^4 N
(E) 4.5×10^4 N

159. A bullet has a mass of 7.5 g. It is fired into a ballistic pendulum. The pendulum's receiving block of wood is 2.5 kg. After the collision, the pendulum swings to a height of 0.1 m. What is the approximate velocity of the bullet?

(A) 50 m/s
(B) 100 m/s
(C) 200 m/s
(D) 250 m/s
(E) 500 m/s

160. **[For Physics C Students Only]** An object is placed in an elastic device and displaced from the equilibrium point. The force as a function of displacement is shown in this graph. When the object is displaced by 2 m, what is the total potential energy that it has?

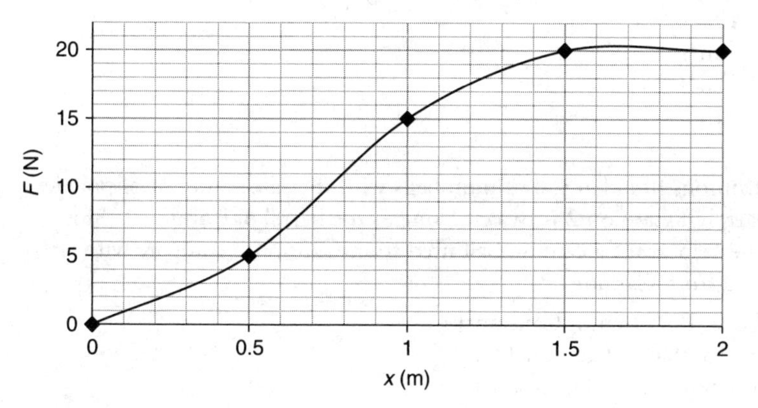

(A) 20 J
(B) 25 J
(C) 30 J
(D) 50 J
(E) 60 J

161. A spring with a constant of 300 N/m is stretched by 0.5 m. What is the force on the spring?

(A) 50 N
(B) 100 N
(C) 150 N
(D) 200 N
(E) 300 N

162. A spring with a constant of 400 N/m is stretched by 0.5 m. What is the potential energy on the spring?

(A) 50 J
(B) 100 J
(C) 150 J
(D) 200 J
(E) 300 J

163. A 1-kg block is attached to a spring with a constant of 100 N/m. The spring is compressed 0.2 m from equilibrium. When the block is let go, what is its velocity as it passes the equilibrium point?

(A) 0.4 m/s
(B) 1.4 m/s
(C) 1.6 m/s
(D) 2.0 m/s
(E) 4.0 m/s

164. Two cannons fire identical cannon balls with the same amount of powder. The barrel of Cannon 2 is twice as long as the barrel of Cannon 1. How does the velocity of a cannon ball fired from Cannon 2 compare with the ball fired from Cannon 1?

(A) Four times that of Cannon 1
(B) Twice that of Cannon 1
(C) 1.4 times that of Cannon 1
(D) One-half that of Cannon 1
(E) The same as Cannon 1

165. A 750-kg aircraft is flying level at 100 m/s. A tailwind blows constantly for 1,200 m and the aircraft's speed increases to 120 m/s. What is the force of the tailwind?

(A) 125 N
(B) 250 N
(C) 2,500 N
(D) 5,000 N
(E) 7,500 N

166. **[For Physics C Students Only]** This graph depicts the potential energy of an object that varies with the distance, *x*. Which of the following statements describes the force on the object?

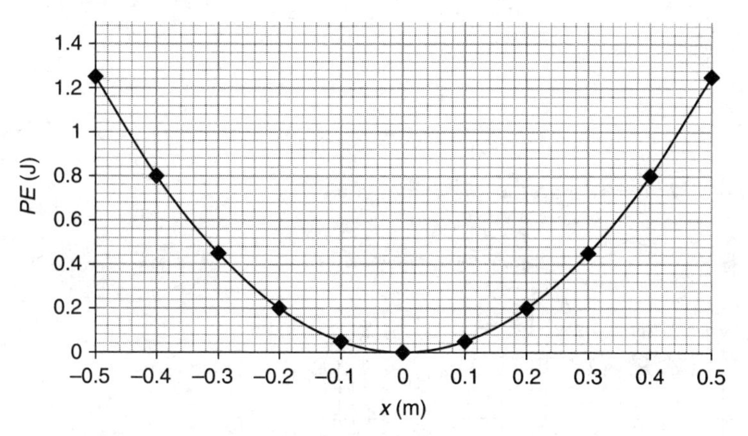

(A) The force is constant.
(B) The force changes, but it is always negative.
(C) The force changes, but it is always positive.
(D) The force changes linearly with a positive slope.
(E) The force changes linearly with a negative slope.

167. A force of 2.0 N exerted on an object for 5.0 m increased the object's velocity by 1.0 m/s. What was the mass of the object?

(A) 0.1 kg
(B) 0.2 kg
(C) 1.0 kg
(D) 20 kg
(E) 200 kg

168. A 70-kg stuntman free falls from a building for 125 m and hits an airbag. The airbag exerts a force on him as he depresses it 5.0 m and comes to a complete stop. What is the force exerted by the air bag?

(A) 1.00×10^2 N
(B) 2.75×10^2 N
(C) 1.75×10^3 N
(D) 2.75×10^3 N
(E) 1.00×10^4 N

169. A man pushes a 100-kg box with a horizontal force of 100 N across a floor for a distance of 60 m in 2 min. What is his power?

(A) 10 W
(B) 20 W
(C) 30 W
(D) 40 W
(E) 50 W

Questions 170–172 are based on the following figure:

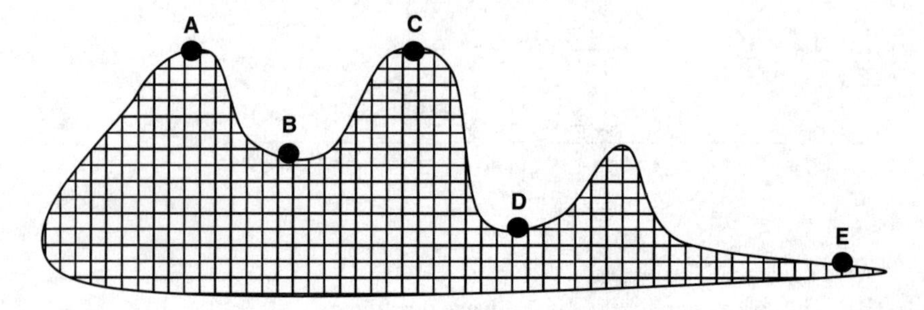

170. The diagram above depicts a roller coaster track with various locations labeled A through E. At which locations will a roller coaster car have the same potential energy?

(A) A and E
(B) B and C
(C) C and D
(D) B and E
(E) A and C

171. In this diagram, at which location will the roller coaster car move the fastest?

(A) A
(B) B
(C) C
(D) D
(E) E

172. For this diagram, at which point will the total energy be the greatest?

(A) A
(B) B
(C) D
(D) E
(E) The total energy is the same at all points

173. A man pushes a lawnmower with a handle at an angle of 60°. He applies a constant 20 N force and moves the lawnmower a horizontal distance of 100 m in 5 min. What is the power of his efforts?

(A) 3 W
(B) 7 W
(C) 10 W
(D) 17 W
(E) 27 W

174. A 2,000-kg airplane flies at 343 m/s. What is its kinetic energy?

(A) 1.18×10^4 J
(B) 1.18×10^5 J
(C) 3.43×10^5 J
(D) 1.18×10^8 J
(E) 3.43×10^8 J

175. A marksman fires a 7.5-g bullet from a 1.2-kg handgun. The bullet travels away at 365 m/s. The bullet strikes a target and loses all its energy as heat (500 J). If the length of the gun barrel is 125 mm, then how much force did the handgun exert on the bullet?

(A) 500 N
(B) 1,000 N
(C) 2,000 N
(D) 4,000 N
(E) 5,000 N

176. Two air cars weighing 500 g each are on a frictionless track. Car A is moving at 0.2 m/s when it collides into Car B, which is at rest. Both cars move away at 0.1 m/s. How much energy is lost as heat during the collision?

(A) 10 percent
(B) 20 percent
(C) 25 percent
(D) 50 percent
(E) 90 percent

177. A rifle fires a 4.0-g bullet at a velocity of 950 m/s. If the length of the rifle barrel is 1.01 m, then what force does the rifle exert on the bullet?

(A) 180 N
(B) 580 N
(C) 1,800 N
(D) 5,800 N
(E) 18,000 N

178. A 2,700-kg cannon fires an 11-kg cannon ball. The length of the cannon's barrel is 244 m. If the powder produces 2.25×10^4 N of force, what is the velocity of the cannonball?

(A) 10 m/s
(B) 50 m/s
(C) 100 m/s
(D) 200 m/s
(E) 500 m/s

179. A 5-kg rubber bowling ball rolls down a 10-m frictionless ramp inclined at 45° as seen in the figure. It rolls out onto a concrete floor. The coefficient of friction between the rubber and concrete is 0.6.

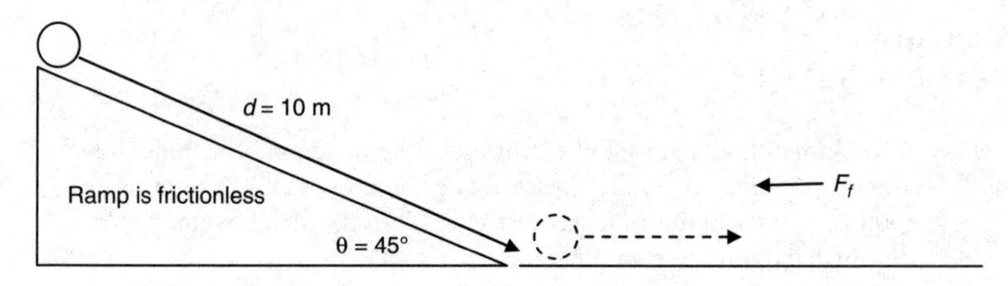

(a) What is the velocity of the ball when it reaches the bottom of the ramp?
(b) What is the kinetic energy of the ball at the bottom of the ramp?
(c) How far does the ball travel before coming to a complete stop?
(d) How long does it take for the ball to roll from the bottom of the ramp until it comes to a complete stop?

180. A tall sailing ship (m = 1.43×10^6 kg) is moving along the ocean as depicted in the figure at an initial velocity of 10 km/h. A wind blows constantly at a 60° angle as shown. The wind blows on the ship over a distance of 1 km. The ship's final velocity is 13 km/h.

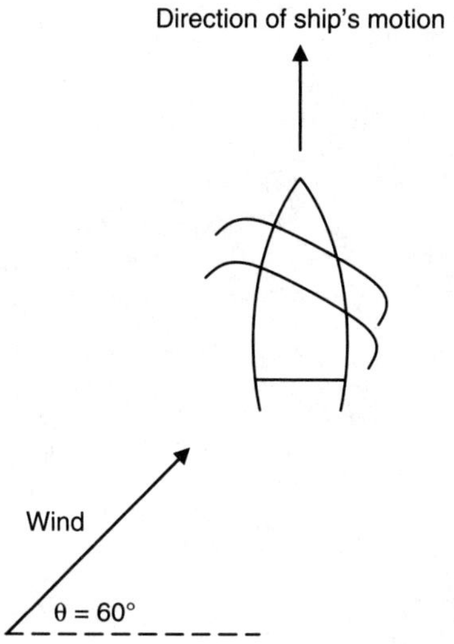

Direction of ship's motion

Wind

$\theta = 60°$

(a) What is the change in velocity of the ship in m/s?
(b) What is the ship's change in kinetic energy?
(c) What is the force of the wind?

CHAPTER 7

Gravitation and Circular Motion

181. A car moves in a horizontal circle with a radius of 10 m. The tangential velocity of the car is 30 m/s. What is the car's acceleration?

(A) 3 m/s^2 toward the center
(B) 3 m/s^2 away from the center
(C) 90 m/s^2 toward the center
(D) 90 m/s^2 away from the center
(E) 270 m/s^2 toward the center

182. If the car in **Question 181** has a mass of 1,000 kg, then what is the force of friction acting on the car?

(A) 3,000 N toward the center
(B) 3,000 N away from the center
(C) 90,000 N away from the center
(D) 90,000 N toward the center
(E) 90,000 N vertically

183. A satellite orbits the Earth at a distance of 100 km. The mass of the satellite is 100 kg, while the mass of the Earth is approximately 6.0×10^{24} kg. The radius of the Earth is approximately 6.4×10^6 m. What is the approximate force of gravity acting on the satellite?

(A) 4×10^4 N
(B) 6.2×10^6 N
(C) 4×10^8 N
(D) 6.2×10^9 N
(E) 4×10^{14} N

184. Two satellites of equal mass orbit a planet. Satellite B orbits at twice the orbital radius of Satellite A. Which of the following statements is true?

(A) The gravitational force on Satellite A is four times less than that on Satellite B.
(B) The gravitational force on Satellite A is two times less than that on Satellite B.
(C) The gravitational force on the satellites is equal.
(D) The gravitational force on Satellite A is two times greater than that on Satellite B.
(E) The gravitational force on Satellite A is four times greater than that on Satellite B.

185. A 70-kg astronaut floats at a distance of 10 m from a 50,000-kg spacecraft. What is the force of attraction between the astronaut and spacecraft?

(A) 2.4×10^{-6} N
(B) 2.4×10^{-5} N
(C) Zero; there is no gravity in space
(D) 2.4×10^{5} N
(E) 2.4×10^{6} N

186. The centripetal acceleration on a 1,000-kg car in a turn is 1×10^5 m/s^2. The radius of the turn is 10 m. What is the car's velocity?

(A) 1×10^1 m/s
(B) 1×10^2 m/s
(C) 1×10^3 m/s
(D) 1×10^4 m/s
(E) 1×10^5 m/s

187. An ice skater skates around a circular rink with a diameter of 20 m. If it takes her 62.8 s to go around the rink once, what is the coefficient of friction of the ice?

(A) 0.01
(B) 0.10
(C) 0.20
(D) 0.30
(E) 0.50

188. A proposed "space elevator" can lift a 1,000-kg payload to an orbit of 150 km above the Earth's surface. The radius of the Earth is 6.4×10^6 m and the Earth's mass is 6×10^{24} kg. What is the gravitational potential energy of the payload when it reaches orbit?

(A) 1.0×10^3 J
(B) 2.7×10^6 J
(C) 6.1×10^{10} J
(D) 2.7×10^{12} J
(E) 1.0×10^{15} J

189. A warrior spins a slingshot in a horizontal circle above his head at a constant velocity. The sling is 1.5 m long and the stone has a mass of 50 g. The tension in the string is 3.3 N. When he releases the sling, what will the stone's velocity be?

(A) 5 m/s
(B) 10 m/s
(C) 25 m/s
(D) 30 m/s
(E) 50 m/s

Questions 190 and 191 are based on the following graph:

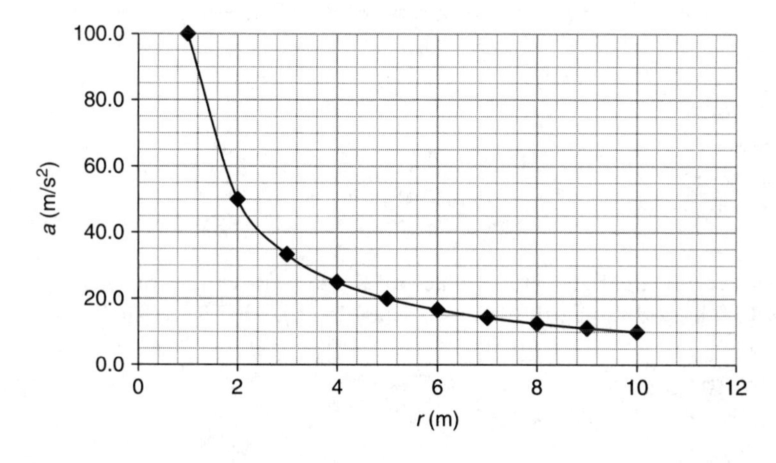

190. Engineers have designed a centrifuge for studying the effects of high gravity environments on plants and animals. This graph shows the results of the relationship between the radius and the centripetal acceleration. If the scientists want to simulate a "3-G environment," then what should the radius of the centrifuge be?

(A) 1 m
(B) 2 m
(C) 3 m
(D) 5 m
(E) 10 m

191. If an astronaut with a mass of 70 kg was placed in that centrifuge with a radius of 5 m, what would be the centripetal force acting on him?

(A) 30 N
(B) 70 N
(C) 140 N
(D) 210 N
(E) 240 N

192. **[For Physics C Students Only]** The Earth is at an average distance of 1 AU from the Sun and has an orbital period of 1 year. Jupiter orbits the Sun at approximately 5 AU. About how long is the orbital period of Jupiter?

(A) 1 year
(B) 2 years
(C) 5 years
(D) 11 years
(E) 125 years

193. A satellite orbits the Earth at a distance of 200 km. If the mass of the Earth is 6×10^{24} kg and the Earth's radius is 6.4×10^{6} m, what is the satellite's velocity?

(A) 1×10^{3} m/s
(B) 3.5×10^{3} m/s
(C) 7.8×10^{3} m/s
(D) 5×10^{6} m/s
(E) 6.1×10^{7} m/s

194. A block with a mass of 30 kg is hanging still from a string. If you place another block with a mass of 10 kg at a distance of 2 m away, what is the gravitational attraction between the two blocks?

(A) 1×10^{-11} N
(B) 5×10^{-10} N
(C) 1×10^{-10} N
(D) 5×10^{-9} N
(E) 1×10^{-9} N

195. A 1,000-kg car experiences a centripetal force of 1.8×10^5 N while making a turn. The car is moving at a constant speed of 30 m/s. What is the radius of the turn?

(A) 0.2 m
(B) 1 m
(C) 2 m
(D) 4 m
(E) 5 m

196. A skater holds out her arms level to the ground as she spins. In one hand she holds a tennis ball. Each of her arms is 1 m long. If the tennis ball travels at 5 m/s, what is its centripetal acceleration?

(A) 5 m/s^2
(B) 10 m/s^2
(C) 15 m/s^2
(D) 20 m/s^2
(E) 25 m/s^2

197. Mars orbits the Sun at a distance of 2.3×10^{11} m. The mass of the Sun is 2×10^{30} kg, and the mass of Mars is 6.4×10^{23} kg. Approximately what is the gravitational force that the Sun exerts on Mars?

(A) 1.6×10^{20} N
(B) 1.6×10^{21} N
(C) 3.7×10^{21} N
(D) 3.7×10^{32} N
(E) 3.7×10^{42} N

198. A record player has four coins at different distances from the center of rotation. Coin A is 1 cm away, Coin B is 2 cm away, Coin C is 4 cm away, and Coin D is 8 cm away. If the player is spinning 45 rotations/min, what coin has the greatest velocity?

(A) Coin A
(B) Coin B
(C) Coin C
(D) Coin D
(E) All the coins have equal velocities

199. A space shuttle is in orbit around the Earth. It travels at an unknown speed (v) at an orbital radius (r). The commander fires the engine and the speed doubles to $3v$. What happens to the orbit if the centripetal acceleration stays the same?

(A) The orbital radius decreases by one-third.
(B) The orbital radius decreases by a factor of nine.
(C) The orbital radius remains the same.
(D) The orbital radius triples.
(E) The orbital radius increases by a factor of nine.

200. This graph depicts the tangential velocities of several circular space stations with different radii. All the stations are spinning. Which of the following statements is true?

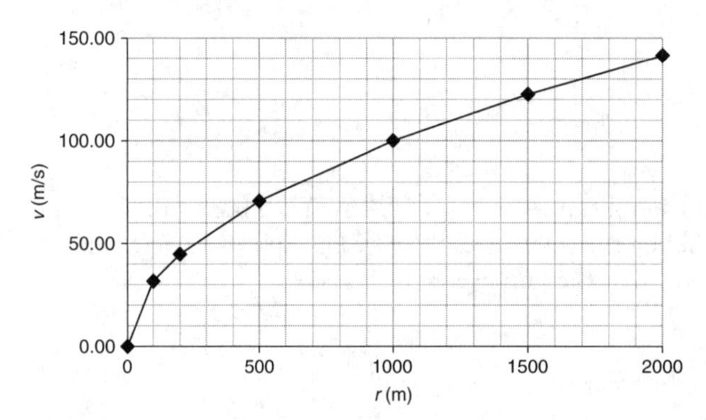

(A) The centripetal accelerations of the short radii space stations are greater than 10 m/s^2; those of the larger ones are less than 10 m/s^2.
(B) The centripetal accelerations of the short radii space stations are greater than 5 m/s^2; those of the larger ones are less than 5 m/s^2.
(C) The centripetal accelerations of all the stations are the same at 5 m/s^2.
(D) The centripetal accelerations of all the stations are the same at 10 m/s^2.
(E) The centripetal accelerations of the short radii space stations are less than 10 m/s^2; those of the larger ones are greater than 10 m/s^2.

201. The Moon has a mass of 7.4×10^{22} kg and a radius of 1.7×10^{6} m. What is the force of gravity experienced by a 70-kg astronaut standing on the lunar surface?

(A) 10 N
(B) 50 N
(C) 100 N
(D) 120 N
(E) 150 N

202. A bicycle wheel has a radius of 0.5 m. When it spins, it completes one full turn in 1.6 s. A pebble wedged in the tread has a mass of 10 g. What is the centripetal force on the pebble?

(A) 0.01 N
(B) 0.08 N
(C) 0.1 N
(D) 0.8 N
(E) 1 N

203. The Moon has a mass of 7.4×10^{22} kg and a distance from the Earth of 3.8×10^{8} m. The Earth's mass is 6×10^{24} kg. What is the gravitational potential energy of the Moon?

(A) 2.0×10^{20} J
(B) 7.8×10^{28} J
(C) 2.0×10^{30} J
(D) 7.8×10^{30} J
(E) 2.0×10^{40} J

204. **[For Physics C Students Only]** Saturn has an orbital period of 29 Earth years. What is its approximate orbital distance?

(A) 1 AU
(B) 2 AU
(C) 5 AU
(D) 10 AU
(E) 15 AU

205. The coefficient of friction between the rubber tires of a car and dry concrete is $\mu = 0.64$. If a car enters a horizontal turn with a radius of 10.0 m, what is the maximum velocity that the car can have and still hold the road?

(A) 4 m/s
(B) 8 m/s
(C) 32 m/s
(D) 64 m/s
(E) 144 m/s

206. A spinning top has a radius of 2 cm. If the top takes 0.06 s to complete one rotation, what is the centripetal acceleration at the edge of the top?

(A) 10 m/s^2
(B) 22 m/s^2
(C) 100 m/s^2
(D) 220 m/s^2
(E) 1,000 m/s^2

207. You swing a 100-g object attached to a 2-m string in a circle above your head. The velocity of the object is 12 m/s. What is the centripetal force on its mass?

(A) 0.72 N
(B) 7.2 N
(C) 72 N
(D) 720 N
(E) 7,200 N

208. Four planets, A through D, orbit the same star. The relative masses and distances from the star for each planet are shown in the table. Which planet has the highest gravitational attraction to the star?

Planet	Relative mass	Relative distance
A	2 m	r
B	m	0.1 r
C	0.5 m	2 r
D	4 m	3 r

(A) Planet A
(B) Planet B
(C) Planet C
(D) Planet D
(E) All have the same gravitational attraction to the star

209. A 1,000-kg satellite orbits the Earth in a circular orbit at an altitude of 1,000 km. The Earth's mass is 6.0×10^{24} kg and its radius is 6.4×10^{6} m.

(a) How does the force of gravity on the satellite compare with the centripetal force on the satellite? What is the magnitude of the force of gravity acting on the satellite?
(b) What is the magnitude of the satellite's tangential velocity?
(c) What is the gravitational potential energy of the satellite?
(d) What is the value of the acceleration due to gravity at this altitude?

210. A 1,000-kg car makes a turn on a banked curve. The radius of the turn is 300 m and the turn is inclined at an angle ($\theta = 30°$). Assume that the turn is frictionless.

(a) Draw a free-body diagram of this situation and label all the forces on the car.
(b) Calculate the car's maximum velocity.
(c) Calculate the centripetal force on the car.

CHAPTER 8

Rotational Motion (For Physics C Students Only)

211. The study of rotational motion uses what coordinate system?

(A) x, y, z
(B) R, θ, Z
(C) R, θ, Φ
(D) x, θ, Φ
(E) z, θ, Φ

212. What is moment of inertia?

(A) The integral of volume
(B) A function of shape
(C) Resistance to rotation
(D) Rotational equivalent of mass
(E) The integral of volume around the z axis

213. How can one tell if a star has a companion or planet system?

(A) By measuring its rotational energy
(B) By calculating its center of mass
(C) By calculating its center of gravity
(D) By measuring its brightness
(E) By triangulating its distance from Earth

214. Why is a person's weight greater at the Earth's equator than at its poles?

(A) This is because the Earth is an oblate spheroid with many mountains near the equator that increase the Earth's mass.
(B) Heat at the equator expands the person.
(C) The Coriolis effect adds to the weight of gravity.
(D) Cold at the poles makes a person eat more.
(E) The radial component of the Earth's acceleration adds a slight amount to the gravitational attraction everywhere but the poles.

215. What is Kepler's first law of planetary motion?

(A) The planets orbit the Sun.
(B) The orbits of the planets are circles with the Sun at the foci.
(C) The orbits of the planets are ellipses with the Sun at one focus.
(D) The orbits of the planets are hyperboloids with the sun at the foci.
(E) The orbits of the planets are all planar.

216. What is the main difference between linear motion and rotational motion?

(A) There are no differences.
(B) One is the analog of the other.
(C) Mass is replaced by the moment of inertia.
(D) Angular momentum is conserved.
(E) The summation of torques is equal to zero.

217. What is the moment of inertia around an axis at the end of a slender bar of mass (m) and length (L) with a large mass (M) of negligible size on its other end? (For a slender rod of mass M, $I = ML^2$ divided by 3.)

(A) $I = ML^2/3 + ML^2$
(B) $I = mL^2/12 + ML^2$
(C) $I = mL^2/2 + ML^2$
(D) $I = mL^2/3 + ML^2$
(E) $I = mL^2/3 + mL^2/12$

218. Where is the center of mass of a binary star system with a white dwarf star of mass (M) and a Sun-like star of mass (m), separated by a distance (L)?

(A) At $R_1 = L \times (1 - m/(m + M))$ from m and $R_2 = m \times L/(m + M)$ from M.
(B) At $R_1 = L^2 \times (1 - m/(m + M))$ from m and $R_2 = m \times L^2/(m + M)$ from M.
(C) At $R_2 = L \times (1 - m/(m + M))$ from m and $R_1 = m \times L/(m + M)$ from M.
(D) At $R_1 = L \times (1 + m/(m + M))$ from m and $R_2 = m \times L/(m + M)$ from M.
(E) At $R_1 = L \times (1 - m/(m - M))$ from m and $R_2 = m \times L/(m - M)$ from M.

219. When a skater performs a spin on ice with his arms outstretched, what happens when he brings his arms close to his body?

(A) His angular acceleration decreases because his moment of inertia was decreased.
(B) His angular acceleration increases because his moment of inertia was decreased.
(C) His angular velocity decreases because his moment of inertia was decreased.
(D) His angular velocity increases because his moment of inertia was decreased.
(E) His angular displacement increases because his moment of inertia was decreased.

220. Why must radians be used in rotational motion problems?

(A) Because radians are the unit of displacement
(B) Because radians are based on the properties of a circle, unlike degrees
(C) Because radians make the numbers come out right
(D) Because radians are based on the properties of a rotation, unlike degrees
(E) Radians are an alternative to degrees; either can be used

221. If a ball attached to a string is being twirled in a circle, what happens to the ball if the string is suddenly cut?

(A) The ball curves away from the circle.
(B) The ball curves into the circle.
(C) The ball continues in a circle because of Newton's first law.
(D) The ball flies off tangent to the circle.
(E) The ball falls to the ground.

222. What is the moment of inertia of a thin-walled cylinder around its central axis?

(A) $I = m \times R^2/2$
(B) $I = m \times R^2$
(C) $I = m \times R^2/12$
(D) $I = 2 \times m \times R^2/3$
(E) $I = m \times (R_{out}^2 - R_{in}^2)$

223. What are the angular velocities of the hour hand and the minute hand of Big Ben in the tower of Westminster Palace in London, England?

(A) 1 rph, 1/12 rph
(B) 2π radians/hr, $\pi/6$ radians/hr
(C) 1 rpm, 0.08 rpm
(D) $\pi/21{,}600$ radians/s, $\pi/1{,}200$ radians/s
(E) 60 ft/hr, 5 ft/hr

224. A piece of space junk has a mass of 3 kg and is orbiting the Earth every 90 min at an altitude of 300 km. The Earth's radius is 6.38×10^6 m. What is the junk's orbital velocity?

(A) 3,710 m/s
(B) 7,420 m/s
(C) 0.001 rad/s
(D) 15,500 m/s
(E) 7,770 m/s

225. What is the kinetic energy of the space junk in **Question 224**?

(A) 9.06×10^5 J
(B) 23,200 J
(C) 15,500 J
(D) 9.06×10^7 J
(E) 1.81×10^8 J

226. Why is torque a vector?

(A) Because it is the product of a force and a length
(B) Because it is the dot product of a force vector and a length vector
(C) Because it is the cross product of a force vector and a position vector
(D) Because the right-hand rule says so
(E) It is not a vector

227. A proposed "space elevator" would carry materials from the ground to an orbit of 300 km. At what speed would the orbiting platform have to move to keep the elevator perpendicular above the Earth platform?

(A) Speed up the space platform to 6 radians/day
(B) Ensure that the elevator tower is rigid
(C) Speed up the Earth platform to 6 radians/day
(D) Slow the space platform down to 6 radians/day
(E) Slow down the Earth platform to 6 radians/day

228. Newton's first law states that a body is in equilibrium if the net forces acting on it are zero. How are torques included in this law?

(A) They are not included in the law.
(B) The summation of forces accounts for the forces causing the torques.
(C) The summation of moments accounts for the torques.
(D) Torques are not forces and do not need to be accounted for.
(E) Torques balance each other out and do not need to be accounted for.

229. What happens to the body on which a torque is acting?

(A) Nothing happens to it.
(B) It causes the body to move.
(C) It causes the body to translate.
(D) It causes the body to rotate.
(E) It causes the body to rotate around the axis perpendicular to the torque.

230. When a body is moving in circular motion, what accelerations can be acting on it?

(A) The acceleration vector
(B) Radial acceleration, tangential acceleration, and angular momentum
(C) Radial acceleration, tangential acceleration, and angular acceleration
(D) Radial acceleration, tangential velocity, and angular acceleration
(E) Radial acceleration, tangential acceleration, and angular velocity

231. A car is rounding a flat unbanked curve. At what speed can it go without being thrown from the curve? (Assume the coefficient of friction between the tires and the road is 0.9, the radius of the curve is 50 m, and the mass of the car is 2,000 kg.)

(A) 107 km/hr
(B) 54 m/s
(C) 23 m/hr
(D) 77 km/hr
(E) 77 m/s

232. A bead slides freely and without friction on a circular wire. If the wire is rotated about a diameter, what will happen to the bead?

(A) The bead will stay in its original position.
(B) If the rotational speed is fast enough, the bead will move to a point halfway up the circular wire.
(C) If the rotational speed is too slow, the bead will move to the bottom of the axis of rotation.
(D) Depending on the rotational displacement, the bead will move up the wire.
(E) The bead will reach the top of the wire loop.

233. A race car is speeding on a circular track. It takes 2 min for the race car to finish one circuit of the 1-km radius track. At what speed is the car moving?

(A) 3.14 m/min
(B) 3.14 mph
(C) 3.14 km/s
(D) 3.14 km/min
(E) 3.14 m/s

234. A roller coaster car is moving around a circular loop of the track. At what point is the car's speed fastest? At what point is the car's speed the slowest? (Neglect friction.)

(A) Any point on the circle, because the velocities are all the same.
(B) At the bottom of the loop it will be moving the fastest and at the top the slowest.
(C) It is moving at a constant speed, so no point will be faster or slower.
(D) The fastest points will be at 0° and 180°, with the slowest at 90° and 270°.
(E) It is not possible to tell because the car is self-propelled.

235. On a moving Ferris wheel, at what point will a rider's weight be greatest? At what point will it be the least?

(A) It will be the greatest at 270° and 90° at the least.
(B) The rider's weight will be the same everywhere on the wheel.
(C) It will be greatest at 180° and 0° at the least.
(D) The rider's weight will not change.
(E) At the top, it will be the greatest and at the bottom the least.

236. An anchor block of mass (m) is released from rest and is falling through the water to the bottom of the ocean 3 km down. It is attached to a winch of radius (R) and mass (M) by a light, flexible, and strong Kevlar cable. Neglecting drag, the cable's weight, and slippage on the winch as well as the stretch of the cable, what is the block's speed when it hits bottom?

(A) $v = \omega^2 \times R$
(B) $v = \alpha \times R$

(C) $v = \sqrt{2gh / \left(1 + \frac{M}{2m}\right)}$

(D) $v = \sqrt{2mgh / \left(1 + \frac{M}{2m}\right)}$

(E) $v = \omega^2 \times R + \alpha \times R$

237. What is the moment of inertia of a solid cylinder around an axis parallel to its central axis but along its outside surface?

(A) $MR^2/2$
(B) MR^2
(C) $MR^2/2 + MR^2$
(D) $MR^2/12$
(E) $MR^2/3 + 2MR^2$

238. How much power is a windmill capable of generating in a steady breeze that causes the blades to rotate at 200 rpm? (Assume each of the mill's three 10-m long blades has a mass of 2,000 kg and neglect all friction losses.)

(A) 2.05×10^6 W
(B) 1.23×10^8 W
(C) 3.26×10^5 W
(D) 6.16×10^6 W
(E) 5.88×10^6 W

239. At a carnival show, one of the booths has a shooting gallery. The idea is to hit the target and knock it over. The guns are loaded with BBs that weigh 0.5 g and have a muzzle velocity of 10 m/s. The targets are 25 cm high and hinged at the bottom. After a target is hit by a BB at 5 cm from its top, what will be its angular velocity just after it is hit? (Assume the target has a moment of inertia of 0.015 kg/m^2.)

240. A 2,000-kg car is rounding a banked curve at 170 km/hr. The curve has a radius of 200 m. At what angle must the curve be banked to prevent the car from flying off the curve? (Assume no friction.)

CHAPTER 9

Simple Harmonic Motion

241. A mass (M) is on a frictionless incline of θ radians and is attached to a spring with spring constant, K. When the mass is pulled down the slope, what will be its period of oscillation when released?

(A) $T = 2\pi\sqrt{M \times \sin\theta / K}$
(B) $T = 2\pi\sqrt{M / K}$
(C) $T = 2\pi\sqrt{M \div (K \times \cos\theta)}$
(D) $T = 2\pi\sqrt{M \ \cos\theta / K}$
(E) $T = \pi \times \sqrt{M \ \cos\theta / K}$

242. A 1-m-long narrow rod is fixed to the ceiling. It has a torsional spring constant of C and a rigid rod of length R. The negligible mass is attached perpendicularly to the narrow rod's end. On the end of the rigid rod is a small mass, M. If the mass and rod are twisted through θ radians, at what period will the mass and rod torsionally oscillate?

(A) $T = 2\pi \times \sqrt{M \div K}$
(B) $T = 2\pi \times \sqrt{M \times R \div C}$
(C) $T = 20.07$ cps
(D) $T = 2\pi \times \sqrt{C \div M \times R}$
(E) $T = 2\pi \times \sqrt{M \times R \div C}$

243. A mass is attached to a spring, which is attached to a pivot on the ceiling. The period of the spring's oscillation is equal to the period of the pendulum's oscillation. Assuming no friction in the system, what happens to the mass when it is moved to one side and released?

(A) The mass moves up and down.
(B) The mass swings from side to side.
(C) The mass swings from side to side and moves up and down.
(D) The mass swings from side to side and then moves up and down, and the motions repeat.
(E) The mass moves up and down and then swings side to side, and the motions repeat.

244. The Moon is approximately 384,000 km from the Earth. The Moon revolves around the Earth once every 27.3 days. What is the frequency of the Moon's motion?

(A) 14,065 km/d
(B) 0.037 rpd
(C) 0.0366300366… rpd
(D) 0.037 rph
(E) 27.3 days

245. A bell is hung in Notre Dame's north tower. The bell and its clapper swing with the same period. When Quasimodo tries to ring the bell, it does not sound. Why?

(A) He is not pulling the bell rope hard enough.
(B) The bell and clapper move as one body.
(C) Because they have the same period, the bell and clapper move as one body.
(D) Quasimodo is deaf.
(E) The clapper cannot reach the bell.

246. In **Question 245**, how can Quasimodo make the bell ring using a clapper?

(A) Use a clapper with a smaller mass on the end so it is out of period with the bell.
(B) Use a clapper with a bigger mass on the end so it is out of period with the bell.
(C) Use a shorter clapper so it is out of period with the bell.
(D) Use a longer clapper so it is out of period with the bell.
(E) Use a longer clapper with a larger mass on the end.

247. Why does putting your legs under the seat of a swing at the bottom of its motion, then outwardly swinging your legs at the top of the motion increase the height of the swing's motion?

(A) The kick at the top of the swing adds energy to the system.
(B) Putting your legs under the swing raises your center of gravity.
(C) The kinetic energy of swinging your legs adds to the energy of the swing.
(D) Raising your legs raises your center of gravity.
(E) Raising your legs adds potential energy to the swing, and kicking them adds kinetic energy.

248. What is special about a Foucault pendulum?

(A) Foucault invented it.
(B) It can be used as a clock.
(C) It moves in one plane and the Earth moves under it.
(D) It is the only demonstration that the Earth rotates.
(E) It is very big.

249. A meter stick is held at one end by a frictionless pivot and is held horizontally at the other end. Neglecting air resistance, how far will the meter stick swing when released?

(A) It will swing around the pivot and back to the starting point.
(B) It will swing just short of horizontal on the other side of the pivot.
(C) It will swing just beyond horizontal on the other side of the pivot.
(D) It will swing to horizontal on the other side of the pivot.
(E) It will drop to vertical.

250. A common office toy consists of five steel balls, each suspended by two strings and each touching the adjacent ball(s). When a ball at the end is raised and then dropped, it hits the adjacent ball and the ball at the other end rises. Why?

(A) The energy in the system must be conserved.
(B) The potential energy of the center balls keeps them in place.
(C) The kinetic energy of the moving ball is transferred through the set of balls, when it hits them, to the only ball that can move, causing it to rise.
(D) The potential energy of the moving ball is transferred through the set of balls to the only ball that can move, causing it to rise.
(E) The center balls are glued together and do not move.

251. Why will a pendulum not oscillate in zero gravity?

(A) The pendulum has no weight in zero gravity.
(B) A pendulum requires gravity as well as the pendulum's mass to create the restoring force.
(C) There is no up or down in zero gravity.
(D) The pendulum would be too far from the Earth to work in zero gravity.
(E) The pendulum requires gravity to work.

252. Will a pendulum swing in an accelerating space ship? What would be its period?

(A) Yes, $2\sqrt{m \div a}$
(B) No
(C) Yes, $2\pi\sqrt{M \div A}$
(B) Yes
(D) Yes, $2\pi\sqrt[3]{L \div A}$
(E) Yes, $2\pi\sqrt{L \div A}$

253. If a ball tied to a string is swung around in a horizontal circle, is it moving periodically?

(A) Yes
(B) No
(C) Yes, but only if it moves like a sine wave
(D) Yes, but only if it moves like a cosine wave
(E) Yes, but only if it makes a harmonic sound

254. If the string on the swing ball above is released, what will be the ball's trajectory?

(A) It will continue to move in a circle.
(B) It will curve away from the center point.
(C) It will move in a straight line.
(D) It will move in a straight line tangential to where it was released.
(E) It will move in a straight line tangential to where it was released and curve downward.

255. Some large oil tankers have an antiroll water tank inside the hull that matches the resonant frequency of the ship's hull. So, when ocean waves hit the ship at the resonant frequency, how does the water tank prevent the ship from capsizing in the waves?

(A) The energy of the waves is used by the water in the tank.
(B) The waves enter the tank and are dampened.
(C) The water tank is 180° out of phase with the ship's hull.
(D) The water tank is 90° out of phase with the ship's hull.
(E) The water in the tank is in phase with the ship's hull.

256. A pendulum has a bob of 28 kg and is 38 cm in diameter. It is hung on a wire that is 67 m long. What is its period and frequency?

(A) 0.06 s and 16.43 cycles/s
(B) 16.43 s and 0.06 cycles/s
(C) 0.608689441 s and 16.42873907 cycles/s
(D) 10.62 s and 0.09 cycles/s
(E) 0.09 s and 10.62 cycles/s

257. A mass of 50 kg is held vertically by a spring on each end of the mass. Both springs have a spring constant of 20 N/m. When set in motion, what is the system's period?

(A) 14.04962946 s
(B) 7.024814731 s
(C) 14.05 s
(D) 7.02 s
(E) 0.14 s

258. A mass of 50 kg is held vertically by two springs, one connected to the other. Each spring has a spring constant of 20 N/m. When set in motion, what is the system's period?

(A) 0.14 s
(B) 14.05 s
(C) 7.02 s
(D) 14.04962946 s
(E) 7.024814731 s

259. A mass of 50 kg is held horizontally on a frictionless surface by two springs, one at each end of the mass. Each spring has a spring constant of 20 N/m. When set in motion, what is the system's period?

(A) 14.04962946 s
(B) 7.024814731 s
(C) 14.05 s
(D) 7.02 s
(E) 0.14 s

260. A mass of 50 kg is held horizontally on a frictionless surface by two springs, one connected to the other. Each spring has a spring constant of 20 N/m. When set in motion, what is the system's period?

(A) 0.14 s
(B) 14.05 s
(C) 7.02 s
(D) 14.04962946 s
(E) 7.024814731 s

261. A 10-kg mass is placed on a frictionless surface and attached to a spring that is attached to a fixed wall. The spring's constant is 20 N/m. When set in motion, what is the system's period? What is the period if the system is held vertically?

(A) 4.4 s and 8.9 s
(B) 8.885765876 s for both
(C) 8.885765876 s and 17.77153175 s
(D) 4.4 s for both
(E) 12.6 s for both

262. A 15-kg mass rests on two springs and is held by a spring attached to the ceiling. The spring constant for each of the bottom two springs is 10 N/m; for the upper spring 25, it is N/m. When set in motion, what is the system's period?

(A) 3.6 s
(B) 7.2 s
(C) 1.8 s
(D) 1.2 s
(E) The mass will not move

263. A mass of 12 kg is hung onto a spring attached to the ceiling. The spring's constant is 19 N/m. How far will the spring stretch when the weight is hung, and what will be the system's period when activated?

(A) 6.2 cm and 15 s
(B) 6.2 m and 5 min
(C) 62 mm and 5 s
(D) 6.2 m and 15.6 s
(E) 6.2 m and 156 s

264. A refrigerator compressor that weighs 8 kg is fixed to three springs on the refrigerator frame. Each has a spring constant of 0.01 N/m. What is the natural frequency of the system?

(A) 0.10 cycles/s
(B) 0.03 cycles/s
(C) 0.8 cycles/s
(D) 1.2 cycles/s
(E) 0.003 cycles/s

265. The pendulum on an old mechanical, weight-driven clock has a period of 3 s. What is the length of the clock's pendulum?

(A) 8 cm
(B) 71 mm
(C) 0.08 s
(D) 0.71 m
(E) 0.08 m

266. A blue light wave vibrates at 6.662×10^{-11} Hz. What is its wavelength?

(A) $4{,}900 \times 10^{-10}$ m
(B) $4{,}500 \times 10^{-10}$ m
(C) $4{,}200 \times 10^{-10}$ m
(D) $4{,}500 \times 10^{10}$ m
(E) $4{,}500 \times 10^{-10}$ m

267. A steel ball with a mass of 100 g is dropped onto a steel plate. The collision is perfectly elastic. From what height must the ball be dropped for the vibrating system to have a period of 2 s?

(A) 100 cm
(B) 19.6 m
(C) 0.1 m
(D) 9.8 m
(E) 4.9 m

268. A pendulum consists of a rigid rod that is 8 cm long with a 100-g mass on one end and a frictionless pivot attached to a plate on the other end. What must be done to the plate when the system is inverted (i.e., mass on the top, plate on bottom) to keep the pendulum oscillating?

(A) Nothing, the pendulum will fall over when it is inverted.
(B) Oscillate the plate side to side opposite the pendulum's swing to keep it upright.
(C) Oscillate the plate vertically in time with the pendulum.
(D) Oscillate the plate vertically with acceleration greater than gravity.
(E) Oscillate the plate vertically at the pendulum's frequency.

269. Consider a homogeneous sphere with a diameter of 12,800 km (approximately the size of the Earth). The sphere has a smooth, straight, frictionless hole bored through the diameter. Neglecting air resistance, will a ball dropped into the hole experience periodic motion? And, if so, what is its period?

270. A spring/mass system is vibrating according to 20 cos ωt. The mass is 10 kg, and the spring constant is 5 cm/N. The amplitude constant is 20 cm. Determine the frequency, period, and maximum and minimum amplitudes of the system.

CHAPTER 10

Thermodynamics

271. A 1.0-m-long section of iron railroad gets warmed by the Sun from 25°C to 40°C. The coefficient of thermal linear expansion for iron is 1.2×10^{-5}°C^{-1}. How much did the length of the iron rail change?

(A) −0.18 cm
(B) −0.18 mm
(C) 0 cm
(D) 0.18 mm
(E) 0.18 cm

272. You have 20 moles of a gas in a 1-m^3 sealed container at 125°C. How many atmospheres of pressure are inside the container?

(A) 0.1 atm
(B) 0.6 atm
(C) 1.0 atm
(D) 1.2 atm
(E) 1.8 atm

273. If you have 2 moles of gas at room temperature (25°C), then what is the internal energy of the gas?

(A) 720 J
(B) 3,600 J
(C) 7,200 J
(D) 36,000 J
(E) 72,000 J

274. The mass of an oxygen atom is 5×10^{-26} kg. What is the average speed of an oxygen molecule at room temperature (25°C)?

(A) 50 m/s
(B) 100 m/s
(C) 250 m/s
(D) 500 m/s
(E) 1,000 m/s

275. In an adiabatic process, 0.5 moles of gas at 1,000 K expand to reach a final temperature of 500 K. How much work was done?

(A) −5,000 J
(B) −3,000 J
(C) 0 J
(D) 3,000 J
(E) 5,000 J

276. An engine converts 5,000 J of thermal energy into 2,500 J of work. What is the efficiency of the engine?

(A) 10 percent
(B) 20 percent
(C) 50 percent
(D) 100 percent
(E) 200 percent

277. An ideal heat engine has an efficiency rate of 20 percent. If the heat reservoir has a temperature of 200°C, then what is the temperature of the heat sink?

(A) 0°C
(B) 50°C
(C) 110°C
(D) 200°C
(E) 300°C

278. If 500 J of work was done on a volume of gas in a container with a moveable piston at a constant atmospheric pressure, what was the change in gas volume?

(A) −0.05 m^3
(B) −0.005 m^3
(C) 0 m^3
(D) 0.005 m^3
(E) 0.05 m^3

279. The pressure, volume, and temperature of a gas were changed from Point A (low temperature) to Point B (high temperature) as shown in this graph. How much work was done by the gas?

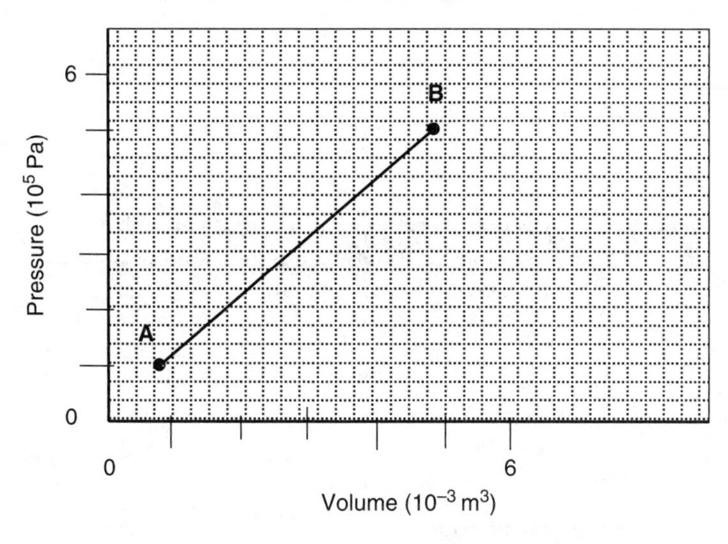

(A) −1,200 J
(B) −120 J
(C) 0 J
(D) 120 J
(E) 1,200 J

280. In which of the following processes is no work done on or by the gas?

(A) Adiabatic
(B) Isobaric
(C) Isothermal
(D) Isochoric
(E) None, all of the processes do work

281. A copper rod has a diameter of 2 cm and a length of 0.5 m. It is placed against a heat source. The temperature of the rod increases from 25°C to 50°C. The thermal conductivity of copper is 390 J/s·m·°C. What is the rate of heat transfer into the copper rod?

(A) 6 J/s
(B) 15 J/s
(C) 2,000 J/s
(D) 2,500 J/s
(E) 3,000 J/s

282. A 20-kg crate slides horizontally on a floor at 0.5 m/s and comes to rest in 25 s. What is the rate of thermal energy transferred between the crate and the floor by friction?

(A) 0.1 W
(B) 0.5 W
(C) 1.0 W
(D) 2.0 W
(E) 10 W

283. One kJ of thermal energy is transferred to a gas in a cylinder with a movable piston. At the same time, 200 J of work is done on the system. What is the change in internal energy of the system?

(A) 5 J
(B) 800 J
(C) 1,000 J
(D) 1,200 J
(E) 5,000 J

284. If 500 J of thermal energy is added to a gas in a cylinder and the temperature remains the same, then which of the following statements is true?

(A) The internal energy increases by 500 J.
(B) The internal energy decreases by 500 J.
(C) The gas does 500 J of work.
(D) 500 J of work is done on the gas.
(E) The pressure inside increases.

285. A 1-m rod of material is heated so that its temperature is increased by 10°C. The rod expands by 0.17 mm. What material is the rod made of?

Material	Coefficient of linear thermal expansion (10^{-5} °C^{-1})
Iron	1.2
Lead	2.9
Copper	1.7
Glass	0.9
Nickel	1.3

(A) Glass
(B) Copper
(C) Iron
(D) Lead
(E) Nickel

286. Two moles of oxygen gas are in a container with a movable piston at 25°C. The pressure is 6,400 Pa. What is the volume of the gas?

(A) 0.001 m^3
(B) 0.01 m^3
(C) 0.1 m^3
(D) 1.0 m^3
(E) 10 m^3

287. One mole of hydrogen gas has an internal energy of 4.3 kJ. What is the temperature of the gas?

(A) 2 K
(B) 16 K
(C) 160 K
(D) 260 K
(E) 360 K

288. A heat engine has a 20 percent efficiency rating. If you put in 2 kJ of thermal energy, how much work will the engine do?

(A) 20 J
(B) 40 J
(C) 200 J
(D) 400 J
(E) 800 J

289. If a heat source for an ideal heat engine has a temperature of 1,000°C and the temperature of the heat sink is 100°C, what is its efficiency?

(A) 7 percent
(B) 21 percent
(C) 42 percent
(D) 64 percent
(E) 70 percent

290. You change the temperature of one mole of nitrogen gas from 25°C to −175°C. What is the change in the gas's internal energy?

(A) −2,400 J
(B) −1,200 J
(C) 0 J
(D) 1,200 J
(E) 2,400 J

291. A concrete slab of a sidewalk heats up in the Sun. The temperature changes from 20°C to 40°C. The coefficient of thermal linear expansion for concrete is $1.2 \times 10^{-5}\,°C^{-1}$. What is the percentage change in the length of the slab?

(A) 0.02 percent
(B) 0.2 percent
(C) 1 percent
(D) 2 percent
(E) 20 percent

292. At a constant pressure, the volume of a gas triples from its original volume. Which of the following statements is true?

(A) The work done on the gas increases by one-third.
(B) The work done by the gas increases by one-third.
(C) The work done by the gas increases threefold.
(D) The work done on the gas increases threefold.
(E) No work is done on or by the gas.

293. A 0.001-m^3 sealed cylinder contains gas at a pressure of 2,500 N/m^2, with a temperature of 25°C. How many gas molecules are in the cylinder?

(A) 3×10^{20}
(B) 6×10^{20}
(C) 3×10^{23}
(D) 6×10^{23}
(E) 1.2×10^{24}

294. A 1.0-m brass rod with a 1-cm radius gets heat transferred to it at a rate of 1 J/s. If the coefficient of linear thermal expansion is 110 $J/s \cdot m \cdot °C$, then what is the change in the rod's temperature?

(A) 3°C
(B) 6°C
(C) 12°C
(D) 18°C
(E) 30°C

295. If the internal energy change during an isochoric process is 1,000 J, how much heat is transferred?

(A) −1,000 J
(B) −100 J
(C) 0 J
(D) 500 J
(E) 1,000 J

296. Assuming a temperature of 125°C for steam, how fast, on average, would a water molecule move?

(A) 75 m/s
(B) 150 m/s
(C) 300 m/s
(D) 600 m/s
(E) 750 m/s

297. You push on the piston of a gas cylinder with a constant force of 2,000 N and the piston adiabatically depresses by 1 m. What was the change in internal energy of the gas?

(A) 100 J
(B) 500 J
(C) 1,000 J
(D) 1,500 J
(E) 2,000 J

298. On which of the following devices must work be done for it to achieve its function?

(A) Steam engine
(B) Electrical motor
(C) Air conditioner
(D) Gasoline engine
(E) All of these devices must have work done on them

299. Ten moles of a gas undergo a pressure and volume change from Point A to Point B as indicated by the arrows in this graph.

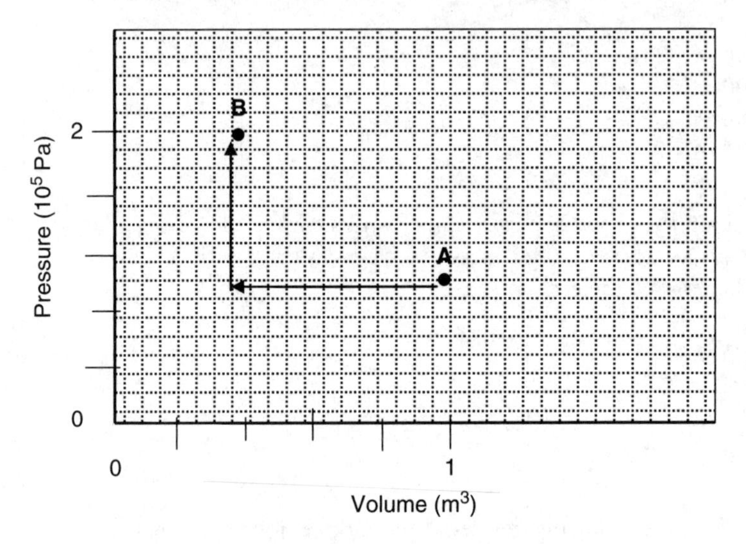

(a) Indicate on the graph how you would calculate the work involved.
(b) Calculate the work involved.
(c) Assuming that the change occurred adiabatically, calculate the change in internal energy.
(d) Calculate the temperature change that occurred.
(e) If the temperature at Point A was 50 K, what was it at Point B?

300. A silver wire has a radius of 0.5 mm and a length of 10 cm. Answer the following questions:

(a) If the wire is heated at one end and the temperature of the wire increases by 200°C, then how much will the wire expand linearly? (The coefficient of linear thermal expansion for silver is 1.9×10^{-5}°C^{-1}.)
(b) What percentage change is the expansion?
(c) If it takes 30 s for the wire to change temperature, how much heat was transferred? (The thermal conductivity for silver is 420 J/[s·m·°C].)

CHAPTER 11

Fluid Mechanics

301. A syringe has an interior diameter of 2 cm and a 20-gauge needle with a nominal inside diameter of 0.603 mm is attached. If the nurse presses on the syringe plunger with 1 N, what will be the pressure in the needle?

(A) 1,100 Pa
(B) 3,180 Pa
(C) 11 Pa
(D) 0.32 Pa
(E) 110 Pa

302. The city water tank is filled from a well. The tank is placed on a 4-m-high hill in the center of the town. The town is mostly at sea level except for the hill. The tank is 25 m tall with the water level maintained by the pump at 24.5 m. What is the water pressure at Mrs. Jones' sink, which has the faucets at 1 m from the ground?

(A) 284,000 Pa
(B) 245,000 Pa
(C) 270,000 Pa
(D) 240,000 Pa
(E) 279,000 Pa

303. A drain valve failed in the town's sewer system, and sewage backed up into the system. The 50-kg, 1-m-diameter manhole covers began popping off. What was the pressure in the sewer system to cause this phenomenon?

(A) 64,000 Pa
(B) 490 Pa
(C) 64 Pa
(D) 620 Pa
(E) 640,000 Pa

304. A glass is filled with water to the brim. Ice is placed in the glass. Will the water overflow, remain the same, or decrease when the ice is added? (Density of water is 9,800 N/m^3 and 9,000 N/m^3 for ice.)

(A) The water will overflow as the ice melts.
(B) The water will decrease.
(C) The water level will remain the same.
(D) The water will overflow.
(E) The water will decrease as the ice melts.

305. Evangelista Torricelli found that his mercury barometer had a column of mercury 760 mm/Hg at his town of Faenza, Italy, which is 50 m above sea level. That same day, he took his barometer into the mountains west of town and found that the mercury column was 720 mm/Hg. When he returned to Faenza, it was raining and the column remained the same. Why? (Assume the temperature remained the same.)

(A) The air density changed.
(B) Some of the mercury evaporated.
(C) A low-pressure front moved into Faenza off the Adriatic Sea.
(D) The lower air density in the mountains caused the mercury column to decrease.
(E) The rain dissolved some of the mercury.

306. In a famous 1654 experiment, Otto von Guericke cast two halves of a 51-cm-diameter sphere. He fit them together so the sphere would not leak, and then he evacuated it. Two teams of eight horses could not pull the hemispheres apart. Later, he tried two teams of 12 horses, but they still could not pull the hemispheres apart. If the sphere was evacuated to 0.1 atmospheres, what force would be required to pull the hemispheres apart?

(A) They cannot be pulled apart by horses
(B) 20,000 N
(C) 18,600 N
(D) 74,555 N
(E) 149,000 N

307. A rock is placed on a boat in a swimming pool. If the rock is taken from the boat and dropped into the pool, what will happen to the water level in the pool?

(A) The water level will rise.
(B) The water level will fall.
(C) The water level will stay the same.
(D) The water level will fall as the rock dissolves.
(E) The water level depends on the size of the rock.

308. Coming home from Greece in the early 1800s, Lady Elgin's ship sank in the Mediterranean with the famous Elgin marbles aboard. British Navy divers searching for the vessel used a weighted 5,000-L wine barrel with an open bottom as a diving bell in which to rest and get air. If the ship was 30 m under water, how much weight must be added to the barrel to sink it to the sunken ship's level? (The density of Mediterranean seawater is 1,029kg/m^3. The volume of wood used in the barrel is ignored, and the air in the barrel is maintained by pumps of 5,000 L.)

(A) 50,400,000 N
(B) 5,040,000 N
(C) 49,000,000 N
(D) 4,900,000 N
(E) 11,334,640 pounds

309. Each of the caryatids on Lady Elgin's sunken ship weighed 50,000 N. The salvage ship displaces 75 m^3 for each 0.5 m on the ship's load line. When all four had been hoisted aboard the salvage ship, how much lower in the water did it ride?

(A) 2.6 cm
(B) 1.3 cm
(C) 19.8 m
(D) 1.3 m
(E) 2.6 m

310. At Barone Dam, the water intake is 7 m in diameter. It narrows to 2 m in the turbine room and increases to 3 m in diameter at the downstream outlet. Using a pitot tube, an engineer measures the outlet flow as 50,000 L/s. What is the water's velocity at each of the three points?

(A) 10, 20, 1 m/s
(B) 7, 16, 1 m/s
(C) 7.07, 16.52, 1.295 m/s
(D) 20, 30, 40 m/s
(E) 10.05, 20.36, 30.56 m/s

311. A 2-ton jack uses a pump to force hydraulic oil into a jack cylinder. With each stroke, the pump moves 5 cc of oil into the 3-cm-diameter jack cylinder. How high does the jack rise with each stroke?

(A) 0.71 cm
(B) 71 cm
(C) 71 mm
(D) 1 cm
(E) 1.07 cm

312. Each of two journal bearings on a ball mill has an area of 2 m^2. The ball mill weighs 270,000 N. At what pressure must oil be injected into a journal bearing to float the ball mill on a film of oil?

(A) 135,000 N/m^2
(B) 140,000 Pa
(C) 150,000 Pa
(D) 540,000 N/m^2
(E) 134,000 Pa

313. There is a 3-m-tall, open water tank at the King Ranch in Texas. It has a rounded outlet at the bottom and is normally capped. The tank is kept filled to 2.75 m by a pump in a nearby well. A ranch hand stops at the tank to get some water and opens the outlet cap. What is the velocity of the water coming out of the outlet?

(A) 7.4 m/s
(B) 54 m/s
(C) 53.9 m/s
(D) 7.3 m/s
(E) 7 m/s

314. If the outlet of the water tank in **Question 313** is 7.5 cm in diameter, will the ranch hand be able to replace the cap on the outlet?

(A) Perhaps, if he has super-human strength
(B) No, because the pressure is too great
(C) Yes, if he is very strong
(D) No, because the force required to replace the cap is too great
(E) No, because the outlet is too small

315. If the tank in **Question 313** was closed and the internal pressure at the outlet was maintained at 690 kPa, what is the outlet velocity when opened?

(A) 1.2 m/s
(B) 37 m/s
(C) 40 m/s
(D) 12 m/s
(E) 20 m/s

316. The Roman aqueducts were built with a slope of approximately 30 cm/km. If the water source was 15 km from Rome, what was the pressure in the fountains fed by that aqueduct?

(A) 0.4 Pa
(B) 0.04 Pa
(C) 0.004 Pa
(D) 4.0 Pa
(E) 40.0 Pa

317. Three beakers are placed on a laboratory bench and are filled with water (density: 9,807 N/m^3), ethyl alcohol (density: 7,740 N/m^3), and sulfuric acid (density: 17,960 N/m^3), respectively. One long glass tube is placed in each of the three beakers, with their open ends in the fluid and the other ends connected to a vacuum pump. Given the density of each fluid, how high will each liquid rise in its tube? (Assume standard air pressure, perfect vacuum, neglect friction, vapor pressure, and capillary action.)

(A) 10.3 mm, 13.1 mm, and 5.6 mm, respectively
(B) The vacuum pump will suck up all the liquids to the same height
(C) The vacuum pump will suck up all the liquids
(D) 10.33 m, 13.1 m, and 5.642 m, respectively
(E) 10.33190 m, 13.091085 m, and 5.641703 m, respectively

318. You are fishing in the local quarry when your friend capsizes the boat while trying to catch a fish. A 45-N bronze fishing trophy, your favorite 225-N maple Captain's chair, a 230-N block of ice that was to be used to cool the fish, and a 180-N steel bar used to stun the fish are now all in the lake. What is the buoyant force exerted on each of the objects? (The densities are as follows: lake water, 9,807 N/m^3; bronze, 858,300 N/m^3; maple, 6,300 N/m^3; ice, 9,040 N/m^3; and steel, 76,500 N/m^3.)

(A) Trophy buoyant force, 5.2 kg; chair buoyant force, 350 kg; ice buoyant force, 250 kg; bar buoyant force, 2.3 kg
(B) Trophy buoyant force, 5.2 N; chair buoyant force, 350 N; ice buoyant force, 250 N; bar buoyant force, 2.3 N
(C) Trophy buoyant force, 5.175 N; chair buoyant force, 350.325 N; ice buoyant force, 249.55 N; bar buoyant force, 2.043 N
(D) Trophy buoyant force, 0.52 N; chair buoyant force, 35 N; ice buoyant force, 25 N; bar buoyant force, 2 N
(E) Trophy buoyant force, 52 N; chair buoyant force, 350 N; ice buoyant force, 250 N; bar buoyant force, 23 N

319. Your mother throws her 20-N cast iron skillet into the quarry lake with a rope attached. What is the buoyant force on the skillet? How much force must be applied to the rope to pull up the skillet? (Cast iron has a density of 70,600 N/m^3.)

(A) 20 N and 20 N
(B) 0.00028 N and 20 N
(C) 0.28 N and 1.72 N
(D) 2.8 N and 17.2 N
(E) 28 N and 172 N

320. Why does a siphon work?

(A) Once filled, head pressure causes the liquid to move through the siphon.
(B) Once filled, air pressure and head pressure cause the liquid to move through the siphon.
(C) Once filled, air pressure causes the liquid to move through the siphon.
(D) Torricelli's law requires it.
(E) Once filled, the intermolecular forces keep the liquid together and then pull the liquid through the siphon.

321. In **Question 320**, how high can the bend in the siphon be above the liquid level and still work?

(A) Only as high as the atmospheric pressure expressed by the liquid
(B) The bore diameter of the tube divided by the pressure head
(C) Only as high as the capillary action of the liquid will allow
(D) Only as high as the pressure head of the liquid
(E) Only as high as the vapor pressure of the liquid

322. How does a toilet work?

(A) Water from the tank pushes the water out of the toilet bowl.
(B) Water from the outlet in the bottom of the toilet bowl pushes the water out of the bowl.
(C) Water from the tank causes the water level to be above the toilet's siphon bend and then it begins siphoning.
(D) Water flowing through the sewer pipe draws the water out of the bowl.
(E) The tank's pressure head causes the water in the bowl to gain velocity and move into the sewer pipe.

323. A fire engine produces 700 kPa in a hose with a 3-cm-nozzle bore. How high can the water from the hose reach? (Neglect friction.)

(A) 72 m
(B) 73 m
(C) 7.3 m
(D) 71 m
(E) 7.1 m

324. Why will a spring-driven watch run faster in the mountains compared with down at the seashore?

(A) Gravity is low in the mountains so the watch runs faster.
(B) Its potential energy is greater in the mountains so the watch runs faster.
(C) Gravity is low in the mountains so the spring's moment of inertia is less and the watch runs faster.
(D) Air in the mountains has less water vapor, the spring moves more easily, and the watch runs faster.
(E) Air in the mountains is less dense, the spring moves more easily, and the watch runs faster.

325. In Philadelphia, two tall buildings are at the corner of Independence Park. When the wind blows from the north, the wind blowing between the two buildings is much faster. Why?

(A) The wind is funneled between the buildings.
(B) The flat area of the park allows the wind to speed up.
(C) Northern winds are faster in Philadelphia.
(D) Because of the continuity of flow, the air moving between the buildings speeds up.
(E) Because of the conservation of energy, the wind must speed up between the buildings.

326. A U tube using mercury shows a height difference of 200 mm. What is the pressure difference between the two sides of the manometer? (Density of mercury is 133,300 N/m^3.)

(A) 27,000,000 Pa
(B) 26,660 Pa
(C) 27,000 Pa
(D) 260,000 Pa
(E) 261,270 Pa

327. A beaker of water is at rest and its water surface is flat. What shapes do the water surface assume when the beaker is rotated around the beaker's vertical axis, then moved horizontally, and then moved vertically?

(A) Semicircle, inclined, and flat
(B) Semicircle, $\pi/4$ angle with the horizontal, and flat
(C) Hyperbola, vertical flat, and lower than when at rest
(D) Parabola, inclined, and flat
(E) Parabola, $\pi/4$ angle with the horizontal, and flat

328. A sailboat heels to port and puts a 25-cm-diameter porthole 1 m under water. What is the average pressure on the porthole? Will the porthole break?

(A) 110,000 Pa; probably
(B) 11,000 $N \cdot m^3$; probably
(C) 11,000 Pa; probably not
(D) 11,000 Pa; probably
(E) 110,000 Pa; probably not

329. A log of cross-sectional Area A is weighted so that it floats vertically in a pond. What is its frequency of oscillation if the friction and the energy imparted to the water are neglected?

330. Given a uniform, open U tube that is partially filled with mercury and free of friction, what is its frequency when the mercury level is caused to oscillate?

CHAPTER 12

Electrostatics

331. An object consists of subatomic particles. It has a net charge of 8.0×10^{-19} C. Which of the following statements is true?

(A) The object only has five protons.
(B) The object has no electrons.
(C) The number of protons and electrons are equal.
(D) The object has five more protons than electrons.
(E) The object has five less protons than electrons.

Questions 332 and 333 refer to the following figure:

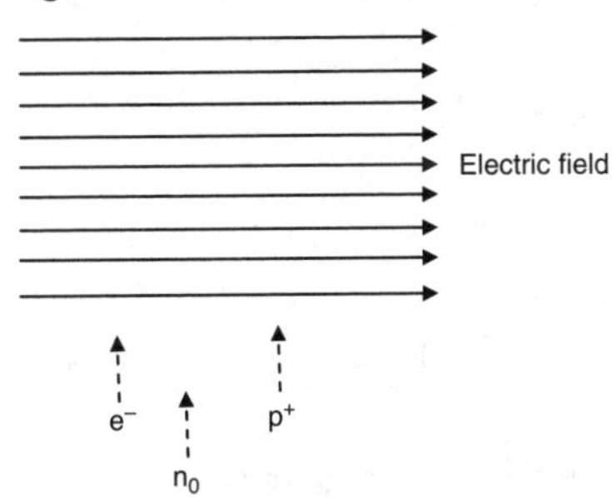

332. An electron, proton, and neutron are fired at a uniform electric field as indicated in the figure. Which of the following statements is true?

(A) The electron will be deflected to the right, the proton will be deflected to the left, and the neutron will pass through the field.
(B) The neutron will be deflected to the right, the proton will be deflected to the left, and the electron will pass through the field.
(C) The electron will be deflected to the left, the proton will be deflected to the right, and the neutron will pass through the field.
(D) The electron will be deflected to the right, the neutron will be deflected to the left, and the proton will pass through the field.
(E) The proton will be deflected to the right, the neutron will be deflected to the left, and the electron will pass through the field.

333. If the electric field in the figure has a strength of 100 N/C, what is the magnitude and direction of the force on the electron?

(A) 1.6×10^{-21} C to the left
(B) 1.6×10^{-21} C to the right
(C) 1.6×10^{-19} C to the right
(D) 1.6×10^{-19} C to the left
(E) 1.6×10^{-17} C to the left

334. An electron is in an electric field surrounding a positive charge. It sits at an equipotential line of 20 V. What happens when it moves to another equipotential line of 40 V if the two lines are 0.5 m apart?

(A) It takes 1.6×10^{-19} J to move from 20 V to 40 V.
(B) The electron does 1.6×10^{-19} J of work moving from 20 V to 40 V.
(C) The electron does 3.2×10^{-19} J of work moving from 20 V to 40 V.
(D) It takes 3.2×10^{-19} J to move from 20 V to 40 V.
(E) It takes 8.0×10^{-20} J to move from 20 V to 40 V.

335. Two protons are 1 μm apart. What is the electric force between them?

(A) 2.3×10^{-22} N attractive
(B) 2.3×10^{-16} N attractive
(C) 0 N
(D) 2.3×10^{-16} N repulsive
(E) 2.3×10^{-22} N repulsive

336. The work done on a 1.0-μC test charge to move it from Point A to Point B in an electric field is -5.0×10^{-5} J. Which of the following statements is true?

(A) The voltage difference is 50 V, and Point B has a higher electrical potential energy (*EPE*) than Point A.
(B) The voltage difference is zero, and both points have the same *EPE*.
(C) The voltage difference is 50 V, and Point A has a higher *EPE* than Point B.
(D) The voltage difference is –50 V, and Point B has a higher *EPE* than Point A.
(E) The voltage difference is –50 V, and Point A has a higher *EPE* than Point B.

Questions 337–340 refer to the following figure:

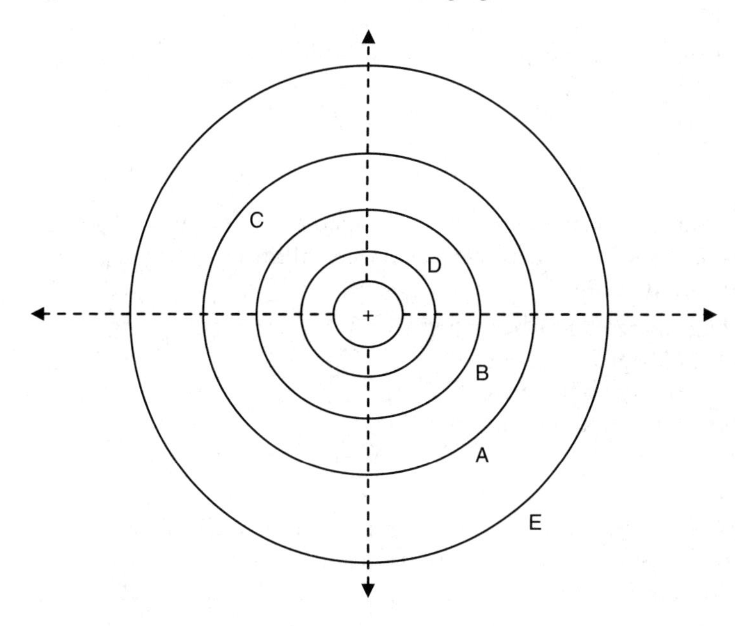

337. In this figure, five negative test charges (A through E) of the same magnitude are shown at different positions in the electric field of a positive point charge (i.e., the solid lines are equipotential). Which charge has the greatest force acting on it?

(A) Test Charge A
(B) Test Charge B
(C) Test Charge C
(D) Test Charge D
(E) Test Charge E

338. In this figure, which charge has the greatest EPE?

(A) Test Charge A
(B) Test Charge B
(C) Test Charge C
(D) Test Charge D
(E) Test Charge E

339. In this figure, which charges have the same *EPE*?

(A) Test Charges A and B
(B) Test Charges A and C
(C) Test Charges C and D
(D) Test Charges B and E
(E) Test Charges A and E

340. Assuming that the distance between each equipotential line in this figure represents the same increment in voltage, which of the following requires the most work to be done?

(A) Moving Charge A to Charge E
(B) Moving Charge E to Charge B
(C) Moving Charge B to Charge C
(D) Moving Charge A to Charge C
(E) Moving Charge D to Charge E

341. Two charged parallel plates are spaced 1 mm apart. A voltage difference of 9 V is placed across them. What is the magnitude of the electric field between the plates?

(A) 0.009 V/m
(B) 0.09 V/m
(C) 9.0 V/m
(D) 90 V/m
(E) 9000 V/m

342. A 20-kg crate slides horizontally on a floor at 0.5 m/s and comes to rest in 25 s. What is the rate of thermal energy transferred between the crate and the floor by friction?

(A) 0.1 W
(B) 0.5 W
(C) 1.0 W
(D) 2.0 W
(E) 10 W

343. How much charge is on each plate of a 1-μF capacitor when a 12.0-V power supply is hooked up to it?

(A) 1.2×10^{-6} C
(B) 1.2×10^{-5} C
(C) 12 C
(D) 1.2×10^{6} C
(E) 1.2×10^{7} C

344. Two parallel plates are separated by 1 mm. The area of the plates is 1 cm^2 and there is nothing but free space between the plates. What is the capacitance?

(A) 8.9×10^{-13} F
(B) 8.9×10^{-12} F
(C) 8.9×10^{-11} F
(D) 8.9×10^{-10} F
(E) 8.9×10^{-9} F

345. **[For Physics C Students Only]** The dotted arrows in this figure depict a uniform electric field. Five wires are depicted by the thick lines labeled A through E. The wires cut through the field. Which of the following represents the electric flux through the wires from least to greatest?

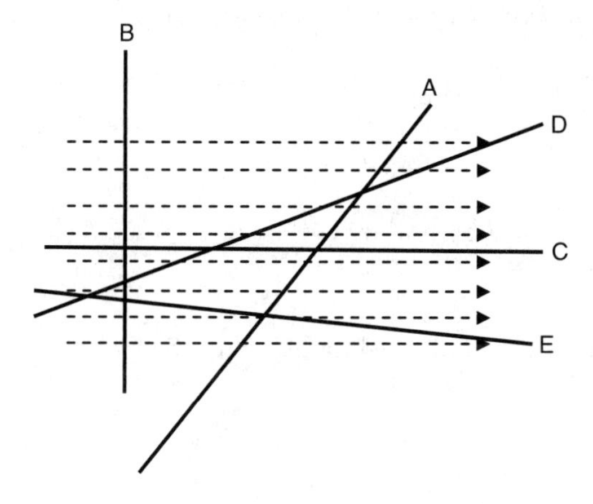

(A) A < B < C < D < E
(B) B < C < A < D < E
(C) C < E < D < A < B
(D) E < C < B < D < A
(E) C < B < D < E < A

346. A 0.1-C charge is located 3 m away from a point charge. What is the electric potential?

(A) 1×10^{-9} V
(B) 3×10^{-9} V
(C) 3×10^{8} V
(D) 1.0×10^{8} V
(E) 1×10^{9} V

347. A 0.2-C charge is in an electric field. It feels like a force of 100 N. What is the strength of the electric field?

(A) 5.0 N/C
(B) 20 N/C
(C) 50 N/C
(D) 200 N/C
(E) 500 N/C

348. A negatively charged object is placed against a neutral object and becomes attracted to one side. Which of the following is true?

(A) The neutral object gains positive charges to become positively charged.
(B) The neutral object loses negative charges to become positively charged.
(C) The neutral object loses positive charges to become negatively charged.
(D) The neutral object gains negative charges to become negatively charged.
(E) Negative charges of the neutral object move to the side opposite the negatively charged object.

349. When placed in a uniform electric field directed to the right, which of the following will move to the right?

(A) Electron
(B) Neutron
(C) Proton
(D) Antiproton
(E) Hydrogen atom

350. When a 0.05-C charge is placed in a field and experiences 100 J of potential energy, what is the voltage?

(A) 5.0 V
(B) 200 V
(C) 500 V
(D) 2,000 V
(E) 20,000 V

351. A test charge is located 10 m away from a point charge in a uniform electric field of 900 N/C. What is the magnitude of the test charge?

(A) 1 μC
(B) 10 μC
(C) 100 μC
(D) 1 mC
(E) 10 mC

352. Two parallel plates have a 120-V potential difference across them and produce a uniform electric field of 1.2×10^6 V/m. How far apart are the plates?

(A) 1 μm
(B) 0.1 mm
(C) 1 mm
(D) 10 mm
(E) 1 cm

353. A 100-μF capacitor is charged by a 1.5-V battery. How much charge is stored on each plate?

(A) 1.5×10^{-6} C
(B) 1.5×10^{-4} C
(C) 0.15 C
(D) 67 C
(E) 150 C

354. What strength and direction of an electric field is necessary to suspend a proton against the force of gravity?

(A) 1×10^6 N/C downward
(B) 1×10^7 N/C upward
(C) 1×10^7 N/C downward
(D) 6×10^7 N/C upward
(E) 6×10^{19} N/C upward

355. A 1-μF capacitor has plates that are 0.2 mm apart. There is only empty space between the plates. What is the area of the plates?

(A) 0.22 m^2
(B) 2.2 m^2
(C) 22 m^2
(D) 220 m^2
(E) 2,200 m^2

356. In this figure, a beam of particles is shot through two electric fields at a target. Both electric fields are uniform and directions are indicated by the arrows. Which of the following particles will reach the target?

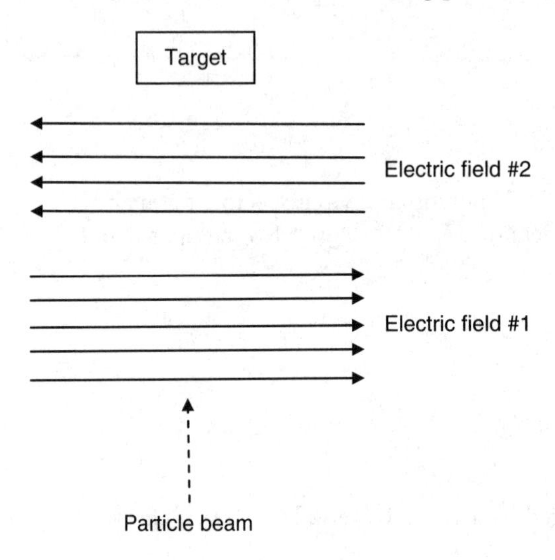

(A) Electrons
(B) Protons
(C) Positrons
(D) Neutrons
(E) Alpha particles

357. A 1-g object with a charge of −0.5 C is at rest. A uniform electric field of 10 N/C is applied to it. The field is oriented to the right. What will be the acceleration of the object in the field?

(A) 50 m/s^2 to the left
(B) 50 m/s^2 to the right
(C) 500 m/s^2 to the right
(D) 500 m/s^2 to the left
(E) 5,000 m/s^2 to the left

358. An electron in an electric field experiences a force of 1×10^6 N opposite the direction of the field. What is the magnitude of the electric field?

(A) 1.7×10^{-25} N/C
(B) 1.7×10^{-24} N/C
(C) 5.9×10^{24} N/C
(D) 5.9×10^{25} N/C
(E) 1.7×10^{26} N/C

359. A proton is initially at rest. A uniform electric field is applied across it by two plates that are 1 m apart. The plates are charged by a 1.5-V battery. The electric field is applied for 1 μs to accelerate the proton uniformly.

(a) What is the strength of the electric field?
(b) What is the force on the proton while the electric field is applied?
(c) What is the acceleration of the proton?
(d) What will be the final velocity of the proton when the field is turned off?
(e) How far will the proton travel during the time that the electric field is applied?

360. Three electrons (A, B, C) are located at 2 nm, 4 nm, and 16 nm, respectively, away from a proton.

(a) What are the magnitudes of the electric fields experienced by each electron? Which one experiences the greatest field?
(b) What are the voltages experienced by each electron? Which one experiences the lowest voltage?
(c) What are the forces experienced by each electron? Which one experiences the greatest force?

CHAPTER 13

Circuits

361. A positive charge of 240 C passes a point in a circuit in 10 s. What is the magnitude of the current?

(A) 0.004 A
(B) 0.04 A
(C) 24 A
(D) 240 A
(E) 2,400 A

362. A 9.0-V battery drives a circuit containing a 10-Ω resistor. What is the current that flows through this circuit?

(A) 0.9 A
(B) 1.1 A
(C) 9.0 A
(D) 11 A
(E) 90 A

363. A 9-V battery drives 3 A of electrical current through a resistor. What is the power dissipated by the resistor?

(A) 0.3 J
(B) 3 J
(C) 6 J
(D) 12 J
(E) 27 J

364. Three resistors (20 Ω, 150 Ω, 500 Ω) are linked in series. What is the equivalent resistance?

(A) 0.001 Ω
(B) 0.06 Ω
(C) 670 Ω
(D) 1.5 kΩ
(E) 1.5 MΩ

Questions 365–368 use the following figure:

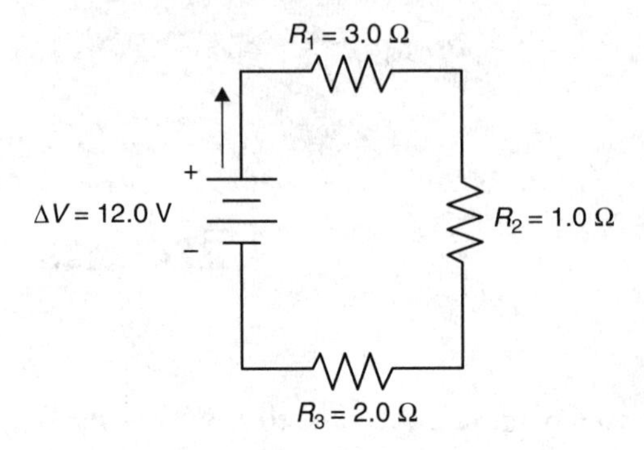

365. What is the current flowing through the circuit shown in the diagram?

(A) 1 A
(B) 2 A
(C) 4 A
(D) 6 A
(E) 12 A

366. Which of the following statements is true about the circuit shown in the diagram?

(A) The voltage drop is greatest across R_1, but R_1 has the least amount of current flowing through it.
(B) The voltage drop is greatest across R_2, but R_2 has the least amount of current flowing through it.
(C) The voltage drop is greatest across R_3, but R_3 has the least amount of current flowing through it.
(D) The voltage drops and current are equal across all resistors.
(E) The voltage drop is greatest across R_1, but the current is equal at all points in the circuit.

367. In this diagram, what is the power dissipated by all of the resistors in the circuit?

(A) 2 W
(B) 6 W
(C) 12 W
(D) 24 W
(E) 48 W

368. In this diagram, what is the voltage drop across the third resistor (R_3)?

(A) 2 V
(B) 3 V
(C) 4 V
(D) 6 V
(E) 12 V

369. Which of the following statements best summarizes a series circuit with three different resistances?

(A) In all parts of the circuit, the resistances are different, the voltage drops are the same, and the current is different.
(B) In all parts of the circuit, the resistances are the same, the voltage drops are the same, and the current is different.
(C) In all parts of the circuit, the resistances are different, the voltage drops are different, and the current is the same.
(D) In all parts of the circuit, the resistances are different, the voltage drops are the same, and the current is the same.
(E) In all parts of the circuit, the resistances are the same, the voltage drops are the same, and the current is the same.

370. When one light in a string of holiday lights goes out, all of the lights go out. Which statement best describes this situation?

(A) All of the lights are wired in parallel.
(B) The lights are wired in a series and in parallel.
(C) Only the parallel portions of the lights went out.
(D) All of the lights are wired in a series.
(E) You cannot tell how the lights are wired.

Questions 371–374 use the following figure:

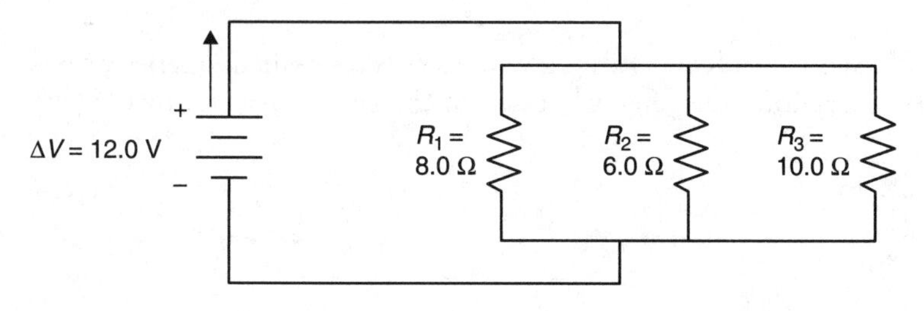

371. For the circuit in the diagram, which of the following expressions will describe the amount of current flowing through the resistors?

(A) $R_1 = R_2 = R_3$
(B) $R_3 > R_2 > R_1$
(C) $R_1 > R_2 < R_3$
(D) $R_2 > R_1 > R_3$
(E) $R_1 < R_2 < R_3$

372. For the circuit in the diagram, what is the equivalent resistance?

(A) 0.04 Ω
(B) 0.4 Ω
(C) 1.0 Ω
(D) 2.6 Ω
(E) 24 Ω

373. For the circuit in the diagram, what is the total current?

(A) 0.5 A
(B) 4.6 A
(C) 12 A
(D) 30 A
(E) 300 A

374. For the circuit in the diagram, how much power is dissipated by the third resistor (R_3)?

(A) 12 W
(B) 14 W
(C) 46 W
(D) 212 W
(E) 300 W

375. **[For Physics C Students Only]** A 100-kΩ resistor is wired in series with a 200-μF capacitor. How long will it take to charge the capacitor to 126 μF?

(A) 1 s
(B) 2 s
(C) 20 s
(D) 1 hr
(E) 5 hr

376. Three capacitors (5 μF, 4 μF, 2 μF) are wired in series with a 9-V battery. What is the equivalent capacitance?

(A) 0.09 μF
(B) 0.95 μF
(C) 1.05 μF
(D) 3.7 μF
(E) 11 μF

377. Three capacitors (5 μF, 4 μF, 2 μF) are wired in parallel with a 9-V battery. What is the equivalent capacitance?

(A) 0.09 μF
(B) 0.95 μF
(C) 1.05 μF
(D) 3.7 μF
(E) 11 μF

378. A 9.0-V battery is hooked up to a 10-μF capacitor. What is the charge on the capacitor?

(A) 9.0×10^{-5} C
(B) 9.0×10^{-4} C
(C) 9.0×10^{-3} C
(D) 9.0×10^{0} C
(E) 9.0×10^{1} C

379. In a circuit, 40 C of charge passes through a 10-Ω resistor in 80 s. What is the voltage that drives the current?

(A) 0.5 V
(B) 1.0 V
(C) 5.0 V
(D) 10 V
(E) 20 V

380. A 100-V power supply is hooked to a resistor. The current flowing through the circuit is 2.0 A. What is the resistance of the circuit?

(A) 0.05 Ω
(B) 0.5 Ω
(C) 2.0 Ω
(D) 5.0 Ω
(E) 50 Ω

381. 100 kW of power is dissipated by a resistor with a 5-A current passing through it. What is the value of the resistance?

(A) 1 kΩ
(B) 2 kΩ
(C) 4 kΩ
(D) 5 kΩ
(E) 201 kΩ

382. A 9.0-V battery is wired in series with three resistors (20 Ω, 30 Ω, 15 Ω). What is the sum of the voltages through the circuit?

(A) 0 V
(B) 2.1 V
(C) 2.7 V
(D) 4.2 V
(E) 9.0 V

383. [**For Physics C Students Only**] A 10.0-μF capacitor is in series with a resistor. When the capacitor discharges to 37 percent of the charge remaining, it takes 50.0 s. What is the value of the resistor?

(A) 5 Ω
(B) 50 Ω
(C) 5 kΩ
(D) 5 MΩ
(E) 50 MΩ

Questions 384–386 use the following figure:

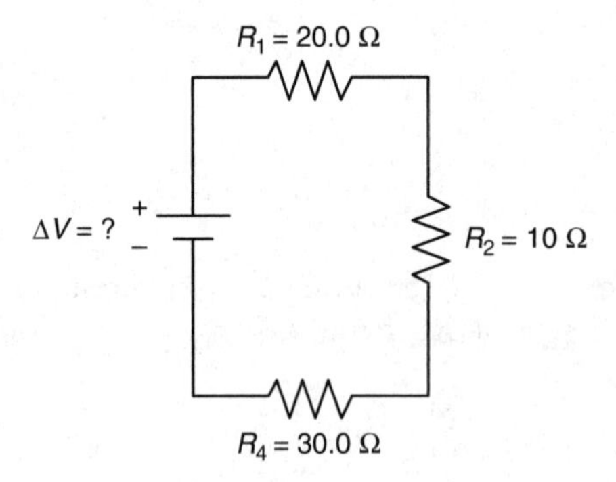

384. For the circuit shown in the figure, what is the voltage of the battery if the current is 2.0 A?

(A) 20 V
(B) 40 V
(C) 50 V
(D) 60 V
(E) 120 V

385. For the circuit shown in the figure, what is the voltage drop across the third resistor if the current is 5.0 A?

(A) 0 V
(B) 50 V
(C) 100 V
(D) 150 V
(E) 300 V

386. For the circuit shown in the figure, what must be the sum of the voltages around the circuit if the current is 10 A?

(A) 0 V
(B) 200 V
(C) 100 V
(D) 300 V
(E) 600 V

Questions 387 and 388 use the following figure:

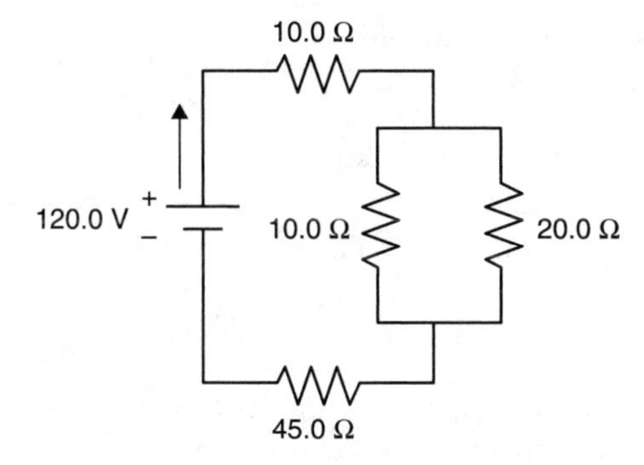

387. For the circuit shown in the diagram, what is the equivalent resistance of the circuit?

(A) 6.7 Ω
(B) 61.7 Ω
(C) 65 Ω
(D) 70 Ω
(E) 85 Ω

388. For the circuit shown in the figure, what is the value of the current leaving the parallel branch of the circuit?

(A) 1.0 A
(B) 1.9 A
(C) 5 A
(D) 10 A
(E) 12 A

389. For the circuit depicted in this figure, find the following:

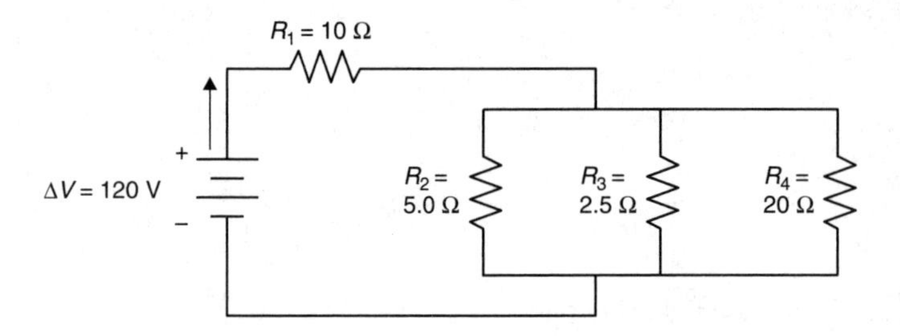

(A) What is the equivalent resistance of the circuit?
(B) What is the current flowing through the circuit?
(C) What is the voltage drop across each resistor?
(D) How much current flows through each resistor?

390. You have a 12-V battery and four 1-Ω resistors. You conduct the following experiments.

Experiment 1

(a) You wire the battery to a resistor, calculate the equivalent resistance, measure the total current, and calculate the power dissipated by the resistor in the circuit.
(b) You add another resistor in series and repeat the measurements and calculations.
(c) You add each resistor in series and repeat the measurements and calculations.
(d) The graphs of each measurement and calculation as a function of the number of resistors in series are shown in the figure.

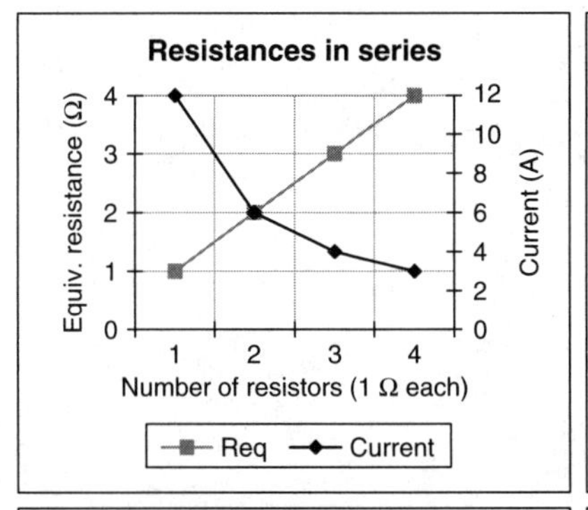

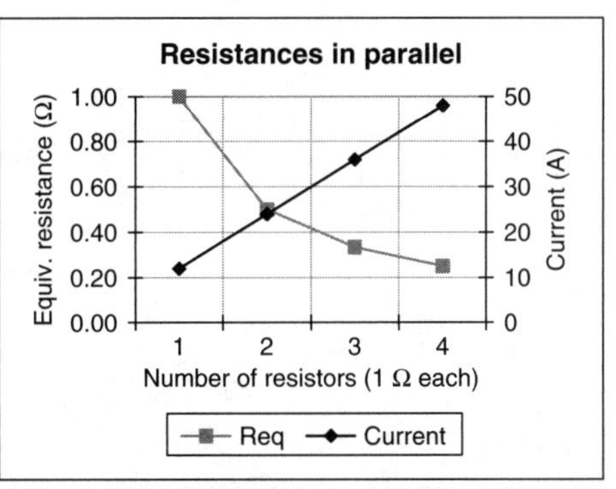

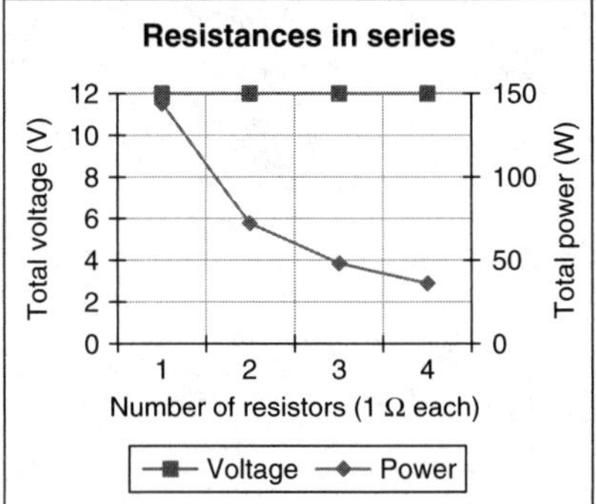

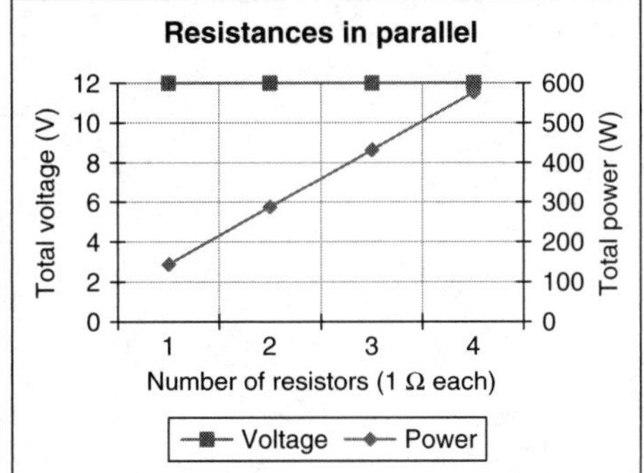

Experiment 2

You follow the procedure for **Experiment 1**, but you wire the resistors in parallel for this second experiment.

Using the graphs, answer the following questions:

(a) Describe the behavior of the equivalent resistance as you add resistors in series versus parallel.

(b) Describe the behavior of the current as you add resistors in series versus parallel.

(c) Describe the behavior of the total voltage as you add resistors in series versus parallel.

(d) Describe the behavior of the power as you add resistors in series versus parallel.

(e) If the resistors were light bulbs in strings of holiday lights, one wired in series and one in parallel, what can you say about the brightness of the bulbs as you increase the number? Why?

CHAPTER 14

Magnetism

391. An electron is moving perpendicular to a 1.0-T magnetic field. The electron has a velocity of 1×10^7 m/s. What is the magnitude of the force acting on the electron?

(A) 1.6×10^{-13} N
(B) 1.6×10^{-12} N
(C) 1.6×10^{-10} N
(D) 1.6×10^{-9} N
(E) 1.6×10^{-8} N

392. Five electrons are moving at the same velocity with respect to a magnetic field, but they are moving at different angles relative to the direction of the magnetic field. Upon which electron does the magnetic field exert the least force?

(A) 0°
(B) 30°
(C) 45°
(D) 60°
(E) 90°

393. A proton moves at a speed of 5.0×10^6 m/s at an angle of 30° relative to a magnetic field. It experiences a force of 4.8×10^{-13} N. What is the strength of the magnetic field?

(A) 0.4 T
(B) 0.8 T
(C) 1.2 T
(D) 2.4 T
(E) 4.8 T

Questions 394–396 use the following figure:

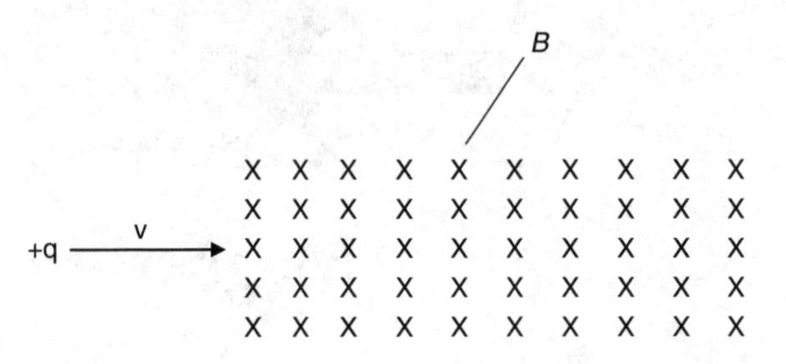

394. A positive charge moves across a magnetic field (see figure). In which direction will the magnetic field exert a force on the charge?

(A) Right
(B) Left
(C) Up
(D) Down
(E) Out of the paper

395. If the charge in the figure above is 1 μC, its velocity is 5,000 m/s, and the strength of the magnetic field is 10 T, then what is the magnitude of the force?

(A) 0.05 N
(B) 0.5 N
(C) 5 N
(D) 500 N
(E) 50,000 N

396. If the charge moving across a magnetic field in the figure was a negative charge instead of a positive one, which of the following changes would occur?

(A) The magnitude of the force would decrease.
(B) The direction of the magnetic field would reverse.
(C) The magnitude of the force would increase.
(D) The magnetic field would exert a force in the opposite direction.
(E) The strength of the magnetic field would decrease.

397. A positive charge travels in a circle within a uniform magnetic field (see figure). The directions of the arrows show the velocity vectors. What is the direction of the force exerted by the magnetic field upon the charge?

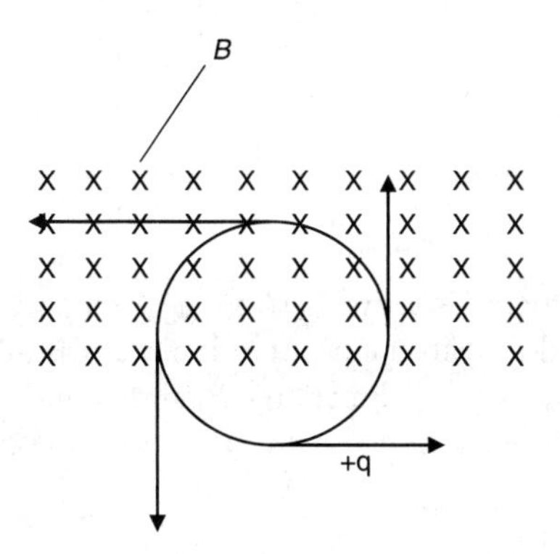

(A) Into the paper
(B) Toward the center of the circle
(C) Away from the center of the circle
(D) Out of the paper
(E) There is no force; all force vectors cancel each other out

Questions 398 and 399 use the following figure:

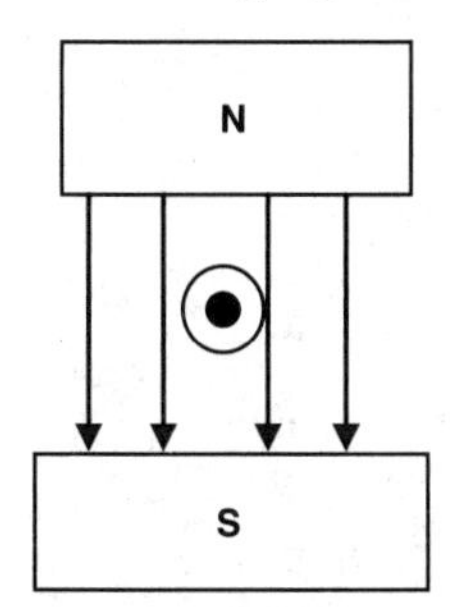

398. In the figure, a piece of wire that is carrying a current passes through a magnetic field. The direction of the current is coming out of the paper. In what direction will the magnetic field exert a force on the wire?

(A) Right
(B) Left
(C) Up
(D) Down
(E) Out of the paper

399. Using the figure, if the current of the wire is 10 A, the strength of the magnetic field is 10 T, and the length of the wire in the field is 1 cm, what is the magnitude of the force exerted on the wire?

(A) 0.01 N
(B) 0.1 N
(C) 1.0 N
(D) 10 N
(E) 100 N

400. This figure is a schematic drawing of a magnetohydrodynamic propulsion unit. The electrodes create a current of ionized seawater as shown by the arrows between them. The direction of seawater and the magnetic field have been indicated. What will be the direction of thrust created by the unit?

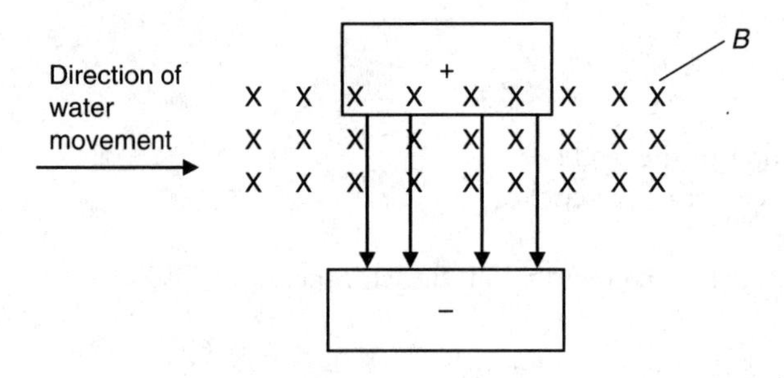

(A) Up
(B) Down
(C) Right
(D) Left
(E) Out of the paper

Questions 401–404 use the following figure:

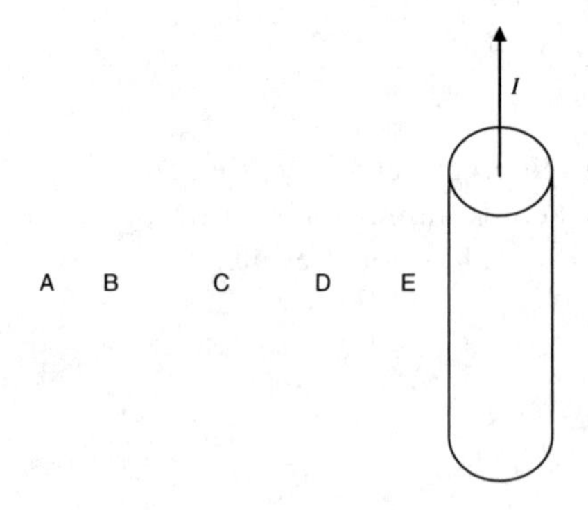

401. The figure shows a section of a wire carrying an electric current (*I*) as indicated. In which direction will the magnetic field generated by the current flow?

(A) Opposite the direction of the electric current
(B) Into the paper to the left of the wire and out of the paper to the right of the wire
(C) Down
(D) In the same direction as the electric current
(E) Out of the paper to the left of the wire and into the paper to the right of the wire

402. The Points A through E on the figure represent various distances from the center of the current-carrying wire. Which point experiences the strongest magnetic field?

(A) Point A
(B) Point B
(C) Point C
(D) Point D
(E) Point E

403. If the current in the wire above is 5.0 A and Point C is 5 mm from the center of the wire, what is the magnitude of the magnetic field at this point?

(A) 2×10^{-7} T
(B) 1×10^{-4} T
(C) 2×10^{-4} T
(D) 1×10^{-3} T
(E) 2×10^{-3} T

404. A positive charge of 1.0 μC moves at a velocity of 5 m/s in the same direction of the electric current at the radius of Point C (5 mm). What is the magnitude of the force exerted by the magnetic field on the charge?

(A) 1×10^{-10} N
(B) 1×10^{-9} N
(C) 2×10^{-9} N
(D) 1×10^{-8} N
(E) 1×10^{-6} N

405. **[For Physics C Students Only]** A solenoid has 100 coils/m of wire carrying a current of 10 A. What is the magnetic field produced by the solenoid?

(A) 1.3×10^{-5} T
(B) 1.3×10^{-4} T
(C) 1.3×10^{-3} T
(D) 1.3×10^{-2} T
(E) 1.3×10^{-1} T

406. A conducting rod ($L = 2.0$ m) is moved along two conducting rails within a magnetic field (B). The rails are hooked up to a light bulb. The resistances of the rails and the rod are negligible, but the light bulb has a resistance of 100 Ω. The rod is moved at a velocity (v) of 10 m/s in the direction indicated. It induces a current (I) in the circuit as indicated in the figure. The magnetic field has a magnitude of 2.0 T and the direction is indicated in the figure. What is the magnitude of the induced electromagnetic field (EMF; ε)?

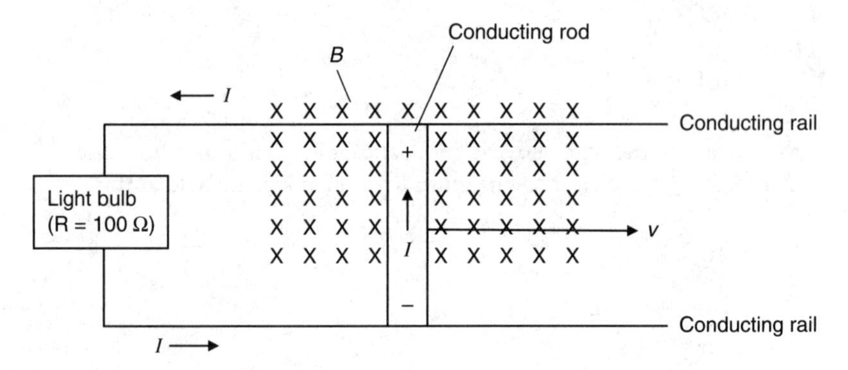

(A) 2 V
(B) 10 V
(C) 20V
(D) 40 V
(E) 100 V

407. A magnetic field of 0.5 T passes through an area of 6.3 cm² that is perpendicular to the field. What is the magnetic flux?

(A) 3.2×10^{-4} Wb
(B) 6.4×10^{-4} Wb
(C) 3.2×10^{-2} Wb
(D) 3.2×10^{-1} Wb
(E) 3.2×10^{0} Wb

408. The primary coil on a transformer has 100 loops. The magnetic flux changes by 10 Wb/s. What is the magnitude of the EMF (ε)?

(A) 1.0 V
(B) 10 V
(C) 100 V
(D) 1,000 V
(E) 10,000 V

409. A magnetic field of 1.0×10^{-4} T is experienced at a point located 5 cm from the center of a current-carrying wire. What is the magnitude of the current in the wire?

(A) 0.025 A
(B) 0.25 A
(C) 2.5 A
(D) 25 A
(E) 250 A

410. Two wires are placed side-by-side as shown in the figure. The wires have currents of the same magnitude, but they are in opposite directions (indicated by the arrows). How will these wires interact?

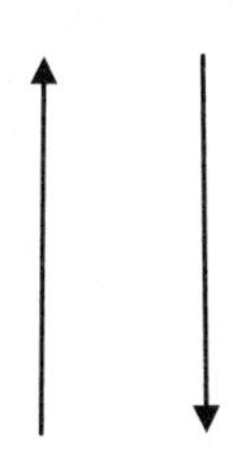

(A) They will attract each other.
(B) They will not affect each other.
(C) They will repel each other.
(D) You can't answer the question without knowing the length of each wire.
(E) You can't answer the question without knowing the value of the current in each wire.

411. A magnetic field exerts a force on a 10-cm length of wire carrying 2 A of current. The magnitude of the force is 0.1 N. What is the strength of the magnetic field?

(A) 0.1 T
(B) 0.5 T
(C) 1.0 T
(D) 2.0 T
(E) 5.0 T

412. A magnetic field ($B = 1.0 \times 10^{10}$ T) exerts a force of 1.6×10^{-3} N on a charge moving at 1×10^{6} m/s perpendicular to the field. What is the magnitude of the charge?

(A) 1.6×10^{-19} C
(B) 1.6×10^{-15} C
(C) 1.6×10^{-9} C
(D) 1.6×10^{-6} C
(E) 1.6×10^{-3} C

413. **[For Physics C Students Only]** A torus has 100 wires/cm wrapped completely around it. The radius of the torus is 0.5 m. The current through the wire is 10 A. What is the strength of the magnetic field at a point that is 1.0 mm from the outside edge of the torus?

(A) 0 T
(B) 4×10^{-6} T
(C) 8×10^{-6} T
(D) 4×10^{-5} T
(E) 8×10^{-5} T

Questions 414–416 use the following figure:

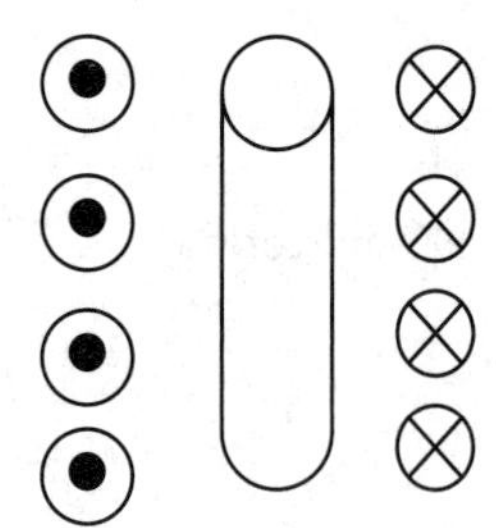

414. The magnetic field produced by a current in a section of current-carrying wire is shown in the figure. What is the direction of the current?

(A) Up
(B) Right
(C) Left
(D) Down
(E) Into the paper

415. If the strength of the magnetic field at 1 mm away from center of the wire is 5×10^{-3} T, then what is the magnitude of the current traveling through the wire?

(A) 0.025 A
(B) 0.25 A
(C) 2.5 A
(D) 25 A
(E) 250 A

416. If a 1-μC charge moves opposite the direction of the current at 1 cm away from the center of the wire and the velocity of the charge is 5 m/s, then what is the magnitude and direction of the force exerted on the charge by the wire's magnetic field?

(A) 2.5×10^{-11} N toward the center of the wire
(B) 2.5×10^{-11} N away from the center of the wire
(C) 2.5×10^{-9} N toward the center of the wire
(D) 2.5×10^{-9} N away from the center of the wire
(E) 2.5×10^{-3} N away from the center of the wire

Questions 417 and 418 use the following figure:

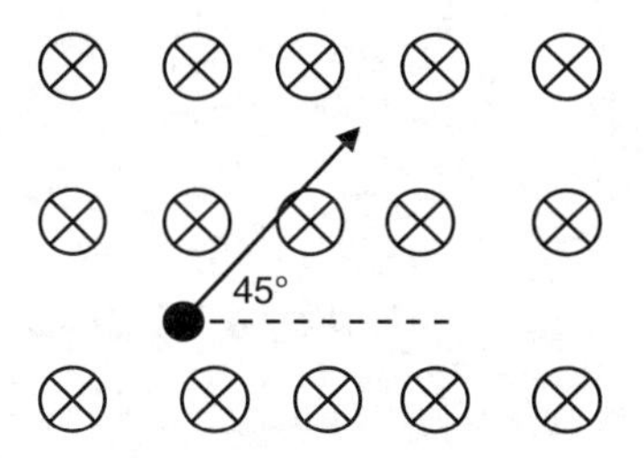

417. A negative charge moves across a magnetic field as shown in the figure. What is the direction of the force exerted upon the charge?

(A) 45°
(B) 90°
(C) 135°
(D) 225°
(E) 315°

418. The magnitude of the charge is $-1\ \mu C$, and it moves at a velocity of 10 m/s in the direction indicated in the figure. If the strength of the magnetic field is 10 T, then what is the magnitude of the force exerted on the charge?

(A) 1×10^{-5} N
(B) 5×10^{-5} N
(C) 7×10^{-5} N
(D) 1×10^{-4} N
(E) 7×10^{-4} N

419. A conducting rod ($L = 2.0$ m) is moved along two conducting rails within a magnetic field (B). The rails are hooked up to a light bulb. The resistances of the rails and rod are negligible, but the light bulb has a resistance of 100 Ω. The rod is moved at a velocity (v) of 5 m/s in the direction indicated. It induces a current (I) in the circuit as indicated in the figure. The magnetic field has a magnitude of 2.0 T and the direction is indicated in the figure.

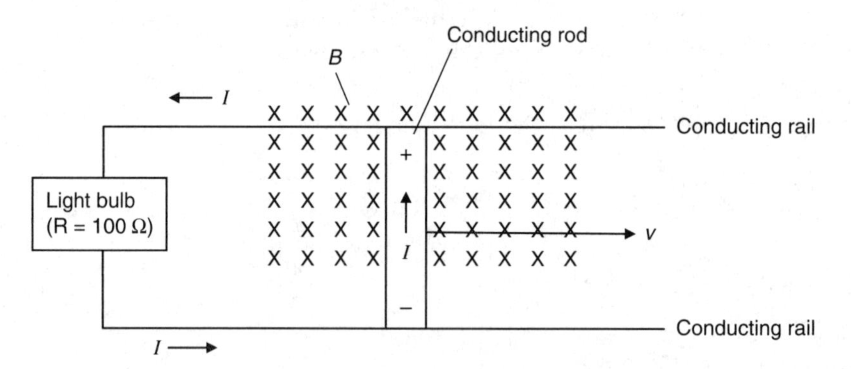

(A) What is the magnitude of the induced EMF (ε)?
(B) What is the current flowing through the circuit?
(C) How much power is dissipated by the light bulb?
(D) How much energy is used by the light bulb in 60 s?

420. [For Physics C Students Only] You have a resistor-inductor circuit that is hooked up in series with a 12-V battery and a switch. The value of the resistor is 10 Ω. The switch is closed at t = 0. The graph shows the current versus time.

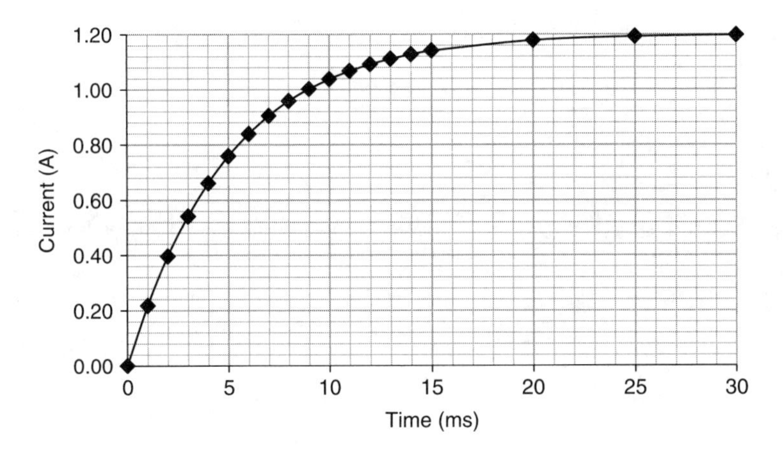

(A) What is the time constant of the circuit?
(B) What is the value of the inductance?
(C) What is the value of the current after two time constants have elapsed?
(D) What is the voltage drop across the resistor after two time constants have elapsed?

CHAPTER 15

Waves

421. Which of the following is an example of a longitudinal wave?

(A) Water wave
(B) Microwave
(C) Sound wave
(D) Radio wave
(E) X-ray

422. The amplitude of a wave is represented by which of the following distances?

(A) Crest to trough
(B) Crest to crest
(C) Trough to trough
(D) Twice the crest
(E) Crest to horizontal axis

423. A wave has a frequency of 100 Hz. What is the period of the wave?

(A) 0.001 s
(B) 0.01 s
(C) 0.1 s
(D) 1 s
(E) 100 s

424. A wave has a frequency of 100 Hz and a wavelength of 1 m. What is the velocity of the wave?

(A) 0.01 m/s
(B) 1 m/s
(C) 10 m/s
(D) 100 m/s
(E) 1,000 m/s

425. The highest sound that a human can hear has a wavelength of 17.2 cm. What is the frequency of this wave?

(A) 20 Hz
(B) 200 Hz
(C) 2,000 Hz
(D) 20 kHz
(E) 20 MHz

426. A jackhammer operator wears a set of protective headphones. Through the headphones, a sound wave is broadcast that is 180° out of phase with the jackhammer sound wave. The result is that he does not hear the sound of the jackhammer. The two sound waves are an example of which of the following?

(A) Standing wave
(B) Transverse wave
(C) Destructive interference
(D) Constructive interference
(E) Doppler effect

427. You have an open tube with a length of 0.5 m. What is the fundamental frequency that can be played in that tube at 20°C?

(A) 34 Hz
(B) 86 Hz
(C) 172 Hz
(D) 343 Hz
(E) 686 Hz

428. A screen is located 2.0 m from a diffraction grating. The spacing within the grating is 0.5 mm. If a light of 475 nm shines through the grating onto the screen, how far will the second-order spot be from the center?

(A) 3.8 μm
(B) 3.8 mm
(C) 3.8 cm
(D) 0.38 m
(E) 3.8 m

429. A wave of green light ($\lambda = 510$ nm) enters a block of material. The wavelength of light within the material is 638 nm. What is the index of refraction of the material?

(A) 0.5
(B) 0.7
(C) 0.8
(D) 1.0
(E) 1.2

430. **[For Physics C Students Only].** Which of Maxwell's equations represents Faraday's law, i.e., a changing magnetic flux through a loop of wire induces an electromagnetic field (EMF)?

(A) Equation 0
(B) Equation 1
(C) Equation 2
(D) Equation 3
(E) Equation 4

Questions 431–433 use the following figure:

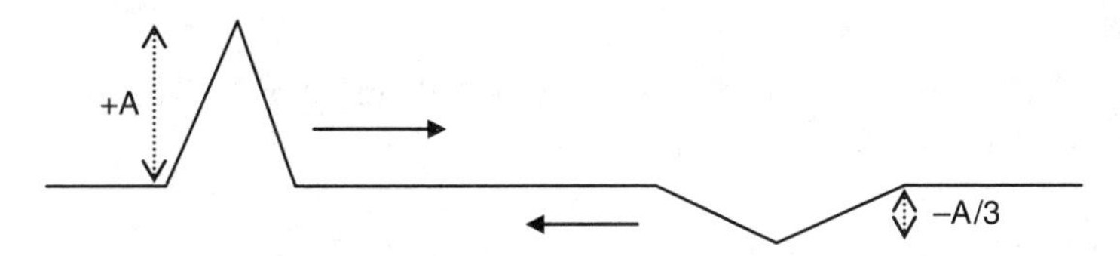

431. Two waves are traveling on a string. The directions and amplitude of each wave is shown in the figure. When the two waves meet, what will be the amplitude of the resulting wave?

(A) −4A/3
(B) −2A/3
(C) 0
(D) 2A/3
(E) 4A/3

432. The figure depicts which of the following phenomena?

(A) Standing wave
(B) Transverse wave
(C) Destructive interference
(D) Constructive interference
(E) Doppler effect

433. After the waves interact, what will happen?

(A) One wave (2A/3) will travel to the right.
(B) One wave (–2A/3) will travel to the left.
(C) There will be no more waves.
(D) One wave (+A) will travel to the right, while one wave (–A/3) will travel to the left.
(E) One wave (–A) will travel to the right, while one wave (+A/3) will travel to the left.

434. A 0.5-m tube is placed in a bucket of water. The tube can be moved up and down to vary the length of the column of air inside. The temperature of both the water and air is 20°C. A 440-Hz tuning fork is struck and placed over the mouth of the tube. The tube is moved up and down until the first resonance can be heard. What is the length of the column of air inside the tube when this happens?

(A) 0.09 m
(B) 0.19 m
(C) 0.27 m
(D) 0.38 m
(E) 0.5 m

435. A tsunami wave travels at 720 km/h and has a period of 10 min. What is the wavelength of the wave?

(A) 2.0 km
(B) 7.5 km
(C) 120 km
(D) 720 km
(E) 750 km

436. [**For Physics C Students Only**] Which of the following will produce a wave of light?

(A) An accelerating neutron
(B) A proton moving at constant velocity
(C) An electron moving at constant velocity
(D) A neutron moving at constant velocity
(E) An accelerating proton

437. The index of refraction of gasoline is 1.4. If a wave of yellow light ($\lambda = 590$ nm) enters the gasoline, what will the wavelength be inside the gasoline?

(A) 245 nm
(B) 421 nm
(C) 590 nm
(D) 826 nm
(E) 1,180 nm

438. A light wave travels from a vacuum into a piece of glass with an index of refraction of 1.5. What is the speed of the wave inside the glass?

(A) 1.5×10^8 m/s
(B) 2.0×10^8 m/s
(C) 2.5×10^8 m/s
(D) 2.75×10^8 m/s
(E) 3.0×10^8 m/s

439. Monochromatic light of wavelength 500 nm passes through a 5-μm slit and produces an interference pattern on a screen. At what angle from the slit would you find the fifth spot from the center?

(A) 10°
(B) 30°
(C) 45°
(D) 60°
(E) 80°

440. Which of the following EMF waves has the longest wavelength?

(A) Ultraviolet light
(B) Radio waves
(C) Blue light
(D) Microwaves
(E) X-rays

441. A favorite radio station is located on the dial at 100 MHz. What is the wavelength of the radio waves emitted from the radio station?

(A) 3 m
(B) 30 m
(C) 300 m
(D) 3 km
(E) 300 km

442. A sound from a moving object changes frequency as it passes you. What is this phenomenon called?

(A) Young's modulus
(B) Maxwell effect
(C) Michelson shift
(D) Doppler effect
(E) Einstein bridge

443. What will be the frequency of the third harmonic in an organ pipe (closed at one end) if the pipe is 3 m long?

(A) 86 Hz
(B) 172 Hz
(C) 343 Hz
(D) 515 Hz
(E) 686 Hz

444. A 440-Hz and a 320-Hz tuning fork are struck simultaneously. What is the beat frequency that you hear?

(A) 80 Hz
(B) 120 Hz
(C) 320 Hz
(D) 380 Hz
(E) 440 Hz

445. A thin film of soap (n = 1.5) is surrounded by air on both sides. A 600-nm beam of light strikes it perpendicular to the surface of the film. What must be the minimum thickness of the soap film to cause constructive interference?

(A) 100 nm
(B) 200 nm
(C) 300 nm
(D) 400 nm
(E) 600 nm

446. Two waves are traveling on a string. The directions and amplitude of each wave is shown in the figure. When the two waves meet, what will be the amplitude of the resulting wave?

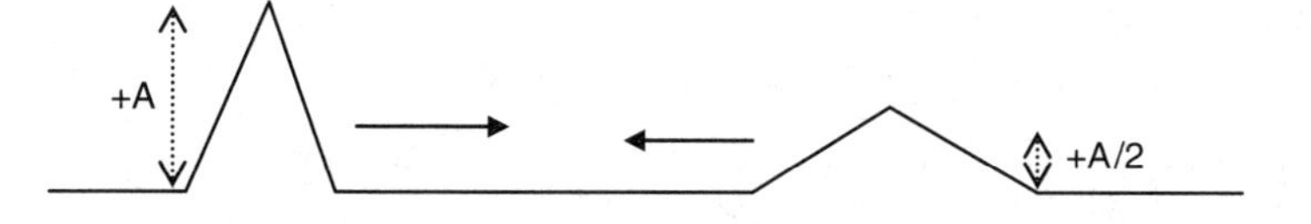

(A) A/2
(B) A
(C) 3A/2
(D) 2A
(E) 4A

447. Which of the following measurements is used to find the wavelength?

(A) Crest to zero displacement
(B) Crest to trough
(C) Trough to zero displacement
(D) Trough to crest
(E) Crest to crest

448. A light wave travels through water at 2.26×10^8 m/s. The temperature of the water is 20°C. What is the index of refraction of the water?

(A) 1.0
(B) 1.1
(C) 1.3
(D) 1.5
(E) 2.0

Use **Experiments 1 and 2** to answer **Question 449**.

Experiment 1

Laser $\lambda = 670$ nm

Distance of screen from slit = 1.0 m

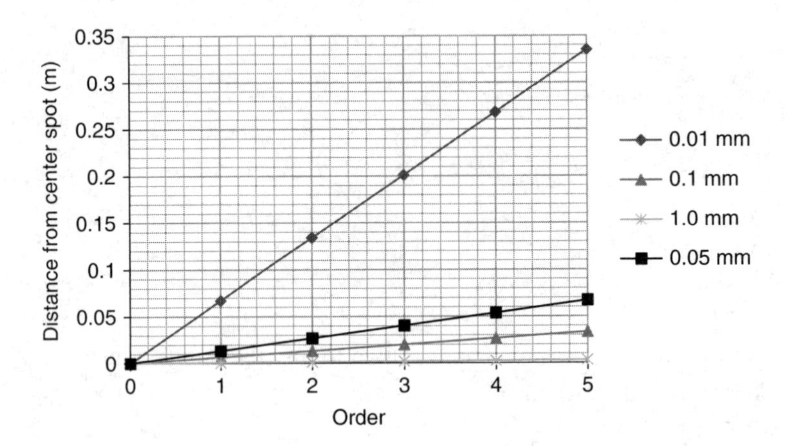

Experiment 2

Slit spacing is 0.01 mm

Distance of screen from slit = 1.0 m

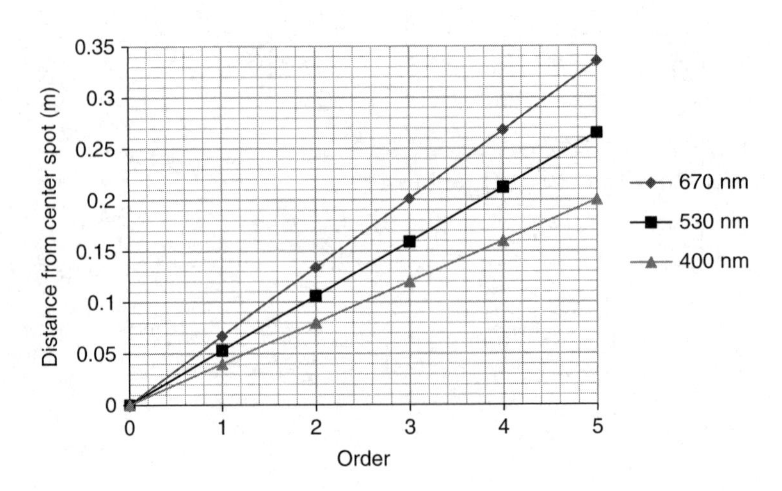

449. A student replicates Young's double slit experiment.

In **Experiment 1**, he uses a 670-nm laser pointer and varies the spacing between the two slits. He measures the distance of each ordered spot from the center and plots that distance versus the order number. In **Experiment 2**, he uses a slit spacing of 0.01 mm and varies the wavelength by using three different laser pointers. He measures the distance of each ordered spot from the center and plots that distance versus the order number.

Using the data, answer the following questions:

(a) How does the distance of the spots from the center change with the order?
(b) What happens to the distance between the spots as you change the slit spacing?
(c) How does the distance between the spots change when you change the wavelength of light?
(d) If you needed to separate the colors from a white light bulb as far as possible, what would be the best way to do it?

450. A thin oil film lies on the top of a glass slide as shown in the figure. When sunlight shines perpendicularly down on the oil film, it appears yellow (blue light – 475 nm is subtracted out). What is the minimum, nonzero, thickness of the film that shows this reflected yellow light? (Show your work.)

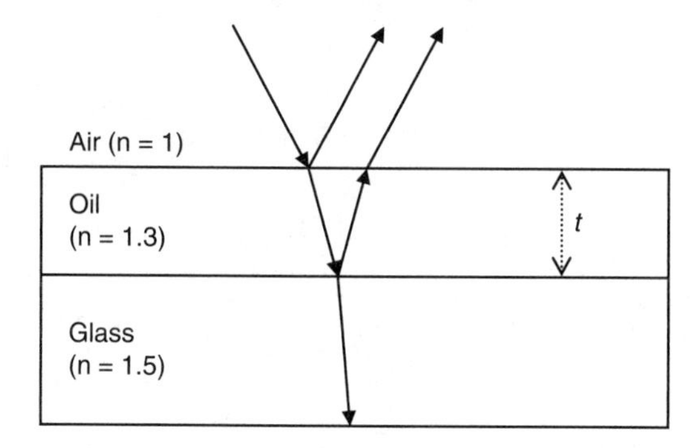

CHAPTER 16

Optics

451. When a beam of light is refracted through a prism, the beam is spread into what colors?

(A) Violet, blue, green, yellow, orange, and red
(B) Blue, violet, green, yellow, red, and orange
(C) Green, yellow, orange, red, blue, and violet
(D) Red, orange, yellow, green, blue, and violet
(E) Orange, red, yellow, green, blue, and violet

452. Why does a piece of glass disappear when immersed in a liquid with the same index of refraction?

(A) The liquid would be dark, and the glass could not be seen.
(B) The liquid and the glass would be the same color.
(C) With the indices the same, light is not refracted, and the glass is not seen.
(D) The glass would be seen at a certain angle.
(E) The liquid causes the glass to melt.

453. What is the focal point of a concave mirror with a radius of 100 cm?

(A) 50 cm
(B) 200 cm
(C) 150 cm
(D) 1 m
(E) 0.5 mm

454. Historians have recorded that Archimedes used mirrors or lenses to burn the Spartan fleet. If he had used mirrors, what would be the focal length of the mirrors Archimedes used to burn the Spartan ships 2 km at sea? Could these be easily made?

(A) 1 km, no
(B) 1 km, yes
(C) 4 km, no
(D) 4 km, yes
(E) 4 m, yes

455. If a person were nearsighted, what shape lens would correct the person's vision?

(A) Concave-convex
(B) Convex-convex
(C) Concave-concave
(D) Convex-concave
(E) Flat-convex

456. To correct for chromatic aberration, what type of lens should be used?

(A) Convex-concave
(B) Two lenses of different refraction
(C) A lens that causes red and blue to refract to the same focal point
(D) A convex-convex lens mated with a concave-concave lens
(E) Concave-convex

457. In a specular reflection, if the incident light beam hits the surface at 45° to normal, at what angle will the reflected beam come from the surface?

(A) 90°
(B) 30°
(C) It will be broken into colors
(D) 45°
(E) 60°

458. Why is it safer to use matte surfaces on surgical instruments rather than polished surfaces during laser surgery?

(A) A matte surface is easier to clean.
(B) A polished surface is easier to clean.
(C) A polished surface would diffuse the laser beam.
(D) A matte surface would diffuse the laser beam.
(E) Laser-safe eyewear should be worn during surgery.

459. An archerfish spits a stream of water at insects above the water surface to knock them down and eat them. If the fish sees an insect at θ degrees under water, how should the fish aim its spit stream to hit the insect?

(A) Much less than θ degrees
(B) Much greater than θ degrees
(C) Slightly greater than θ degrees
(D) Slightly less than θ degrees
(E) At θ degrees

460. To keep as much of the laser beam energy inside a laser fiber, a coating is sometimes placed on the fiber to produce total reflection in the fiber. What refractive index should the coating have?

(A) It should be opaque.
(B) It should be mirrored.
(C) It should be higher than that of the fiber.
(D) It should be lower than that of the fiber.
(E) It should have a critical angle of greater than 41°.

461. A totally reflecting prism can be used to do what to a beam of light?

(A) Reflect the beam 90°
(B) Reflect the beam 180°
(C) Invert the beam
(D) All of the above
(E) Only (A) and (B)

462. Galileo made a telescope to study the heavens. The telescope has a 37-mm diameter plano-convex objective lens with a focal length of 980 mm. According to Galileo's writings, the original eyepiece, which was lost, was a plano-concave lens with a diameter of approximately 22 mm and a focal length of approximately 50 mm. What was its magnification and how did the image appear?

(A) 26.5, upright
(B) 1.7, inverted
(C) 19.6, inverted
(D) 1.7, upright
(E) 19.6, upright

463. Newton's telescope was a reflecting telescope with a spherical mirror and a spherical eyepiece lens. Why was it better than Galileo's telescope?

(A) It eliminated chromatic aberration.
(B) It eliminated spherical aberration.
(C) It had a wider field of view.
(D) Only (A) and (B).
(E) Only (A) and (C).

464. The Hubble Space Telescope is a reflecting telescope. Why was it placed in space?

(A) Because of its size
(B) Because of its focal length
(C) Because of the clarity of the image
(D) Because of the lack of atmospheric aberration
(E) Because of its expense

465. The Cerro Paranal telescope array in southern Chile uses four 8.2-m diameter reflecting telescopes. The site is in the Andes Mountains at 2,600 m above sea level. It is approximately 120 km from the nearest town of any significant size. Why is this a good site for a light telescope? Why isn't it?

(A) Light pollution is low, but you can only view the Southern Hemisphere.
(B) There is plenty of room on the mountains, but it is not easily accessible for most astronomers.
(C) It is above most of the atmosphere, but it is not easily accessible for most astronomers.
(D) It is above most of the atmosphere and has low light pollution, but you can only view the Southern Hemisphere.
(E) It is above most of the atmosphere and has low light pollution, but it is affected by the nearby ocean's humidity.

466. Which color is refracted most when shone on water?

(A) Red
(B) Yellow
(C) Blue
(D) Green
(E) Violet

467. If you stand centered between two mirrors that are 2 m apart, how far away will your image appear in the mirror? How many images will you see?

(A) 1 m, an infinity
(B) 2 m, very many
(C) 2 m, an infinity
(D) 1 m, very many
(E) 2 m, one

468. A spherical concave mirror has a focal length of 200 cm. If it is used for stargazing, how far from the mirror will the star images appear?

(A) 400 cm
(B) 200 cm
(C) 100 cm
(D) 201 cm
(E) 401 cm

469. What kind of lens should be used to correct eyes with presbyopia (i.e., a condition in which the lens of the eye is unable to focus)?

(A) Concave-convex
(B) Convex-convex
(C) Concave-concave
(D) Convex-concave
(E) Flat-convex

470. A person with astigmatism has an eye lens that is not spherical. How can this be corrected?

(A) With a concave-convex lens
(B) With a convex-convex lens
(C) With a concave-concave lens
(D) With a convex-concave lens
(E) With a cylindrical lens

471. How does a "one-way" mirror work?

(A) It has a large index of refraction.
(B) It is partially silvered on one side.
(C) One side is brightly lit.
(D) One side is darkened.
(E) All of the above.

472. After watering a plant, brown spots may sometimes appear on its leaves. Why?

(A) Acid rain
(B) Insects attracted to the water burrow into the leaf
(C) Drops of water act like lenses and focus the sun on the leaf
(D) Dust in the water blocks the sun and the area under the drop dies
(E) The leaf cannot breathe, and the area under the drop dies

473. Why do a cat's eyes glow in the dark?

(A) A layer of phosphorescent material lines the cat's retinas
(B) A cat's eyes are mirrored
(C) All eyes reflect light
(D) A cat's eyes are prismatic
(E) All of the above, except (C)

474. A person looks over the rim of an empty cylindrical cup and sees the far edge of the bottom. While doing this, the person's line of sight forms an angle, α, with the side of the cup. When the cup is filled with water, the person can see the center of the cup from the same angle. If the cup is 6 cm in diameter, how deep is the cup? (The index of refraction for water is 1.333 and 1.000 for air.)

475. A convex-convex lens has a focal length of 20 cm. Determine the image distances when an object is placed at 60 cm, 40 cm, 20cm, 10 cm, and 5 cm from the lens.

CHAPTER 17

Atomic and Nuclear Physics

476. The isotope $_{98}$californium^{251} has a critical mass of 5 kg and a density of 15.1 g/cm^3. What is the diameter of a sphere of californium at critical mass?

(A) 0.43 cm
(B) 4.3 cm
(C) 8.5 cm
(D) 0.85 m
(E) 0.85 mm

477. As of June 2011, how many radioactive elements are naturally occurring? How many are manmade?

(A) 12 and 25
(B) 12 and 37
(C) 28 and 12
(D) 28 and 37
(E) 1 and 36

478. What are the nuclear decay processes?

(A) Alpha, beta, and delta
(B) Alpha, beta, and gamma
(C) Delta, omega, and gamma
(D) Phi, theta, and kappa
(E) Alpha, beta, and chi

479. What particles are emitted during the nuclear decay processes?

(A) A helium atom, an electron, and a photon
(B) A helium nucleus, an electron, and a photon
(C) A quark, an electron, and a photon
(D) A quark, a positron, and a photon
(E) A helium nucleus, an antielectron, and an X-ray

480. If 5 kg of californium were allowed to fission completely, how much energy would be released?

(A) 1,498,962,291 Nm
(B) 1.4989×10^{-8} J
(C) 1,498,962,290 J
(D) 15×10^{7} Nm
(E) 15×10^{8} J

481. How many electrons, protons, and neutrons are in an atom of $_{92}\text{uranium}^{235}$?

(A) 235, 235, 92
(B) 92, 92, 92
(C) 92, 29, 146
(D) 92, 92, 143
(E) 92, 92, 141

482. What governs the energy contained in a photon?

(A) Planck's constant
(B) The speed of light
(C) Its frequency
(D) Its wavelength
(E) All of the above

483. Thorium-234 has a half-life of 24 days and decays through beta emission to protactinium-234. If you were given 5 kg of thorium, how long would it take for it to decay to 2.5 kg?

(A) 72 days
(B) 48 days
(C) 24 days
(D) 12 days
(E) 6 days

484. The element $_{92}\text{uranium}^{238}$ decays to a stable isotope of $_{82}\text{lead}^{206}$. How many alpha particles must be emitted during the decay process?

(A) 8
(B) 10
(C) 32
(D) 40
(E) 16

485. A helium neon laser emits a red wavelength of 614 nm. How much energy is contained in each photon released?

(A) 2 eV
(B) 2.02 eV
(C) 2.0195 4 eV
(D) 2.01 eV
(E) 2.0 eV

486. Why can a slow neutron trigger nuclear fission in uranium^{235} but a slow proton cannot?

(A) Neither can trigger fission.
(B) Both can trigger fission.
(C) The slow proton will only capture an electron.
(D) The slow neutron has no charge, and it can hit the nucleus to cause fission.
(E) Fission is not caused by slow neutrons.

487. Why aren't all the atomic mass units (amu) listed on the periodic table integers?

(A) Some are
(B) Mass of electrons account for the fractional amus
(C) They all should be
(D) Because the amu is the average of all of an element's isotopes
(E) Because some elements have decayed since the start of the universe

488. When an electron and a position collide, what is emitted?

(A) Energy
(B) An alpha particle
(C) A photon
(D) An electron
(E) Two photons

489. Rutherford bombarded $_7$nitrogen^{14} with alpha particles from natural sources to produce oxygen and hydrogen. What were the resulting atomic numbers and atomic mass units?

(A) $_8$Oxygen^{15} and $_1$Hydrogen^3
(B) $_8$Oxygen^{17} and $_1$Hydrogen^1
(C) $_8$Oxygen^{16} and $_1$Hydrogen^2
(D) $_8$Oxygen^{16} and $_1$Hydrogen^1
(E) $_8$Oxygen^{18} and $_1$Hydrogen^1

490. How much energy is contained in photons with the following wavelengths: 2.02×10^{-2} m (microwave), 10.6 mm (mid-infrared), 1.68×10^{-8} m (ultraviolet), and 2.02×10^{-10} m (X-ray)?

(A) 2.02×10^{-7} eV, 1.06×10^{-9} eV, 1.68×10^{-13} eV, and 2.02×10^{-15} eV
(B) 6.13×10^{-7} eV, 1.16×10^{-9} eV, 7.38×10^{-13} eV, and 6.13×10^{-15} eV
(C) 2.02×10^{-9} eV, 1.06×10^{-11} eV, 1.68×10^{-15} eV, and 2.02×10^{-17} eV
(D) 6.13×10^{-9} eV, 1.16×10^{-11} eV, 7.38×10^{-15} eV, and 6.13×10^{-17} eV
(E) 6.13×10^{-11} eV, 1.16×10^{-13} eV, 7.38×10^{-17} eV, and 6.13×10^{-19} eV

491. If uranium^{235} radiates 200 MeV per fission, what is the heat of fission in terms of J/kg?

(A) 6.02×10^{19} J/kg
(B) 8.200×10^{19} J/kg
(C) 8.2×10^{14} J/kg
(D) 8.20×10^{13} J/kg
(E) 6.02×10^{23} J/kg

492. In a cyclotron, an electron is excited to 1 eV. How fast is it moving?

(A) 3.51×10^{11} m/s
(B) 5.93×10^{5} m/s
(C) 3.00×10^{8} m/s
(D) 5.93×10^{8} m/s
(E) 6×10^{6} m/s

493. Because a photon does not have a mass, how can its impact with an atom have an effect?

(A) A photon has a very tiny mass, which allows the effect.
(B) A photon is a quantum of energy, which allows the effect.
(C) Because a photon is wavelike, the wave affects the atom.
(D) Because an atom absorbs photons.
(E) Photons only interact with photons.

494. Why is the photoelectric effect most important?

(A) It allowed Einstein to write a paper.
(B) It showed that light affects atoms.
(C) It helped define the quantum effect of energy.
(D) It led to the development of photocells.
(E) It became a new electric circuit element.

495. Does light behave like a wave or a particle?

(A) It behaves like a wave.
(B) It behaves like a particle.
(C) It depends on the property being examined.
(D) It behaves like a wave and a particle.
(E) It does not behave like a wave or a particle.

496. In nuclear fission, what causes the fission? How can it be regulated?

(A) Particles from split atoms cause fission. Once started, it cannot be stopped.
(B) Radioactive elements are always undergoing fission.
(C) Radioactive elements are always undergoing fission. This can be regulated by keeping the mass below the critical mass.
(D) Neutrons hitting atoms cause fissions. These can be regulated by neutron absorbers.
(E) Photons cause electrons to split atoms. Protecting the atoms from light regulates fission.

497. Do all elements undergo alpha decay?

(A) Yes
(B) No
(C) Only radioactive elements
(D) Only elements with sufficient mass
(E) All elements except hydrogen

498. What happens in beta decay?

(A) The atom's energy changes, causing changes in the nucleus.
(B) A nuclear particle is changed, and an electron or positron is emitted.
(C) The atom's charge changes.
(D) A nuclear particle is changed, and an electron is emitted.
(E) A nuclear particle is changed, and an electron or positron is emitted along with a neutrino and an antineutrino.

499. The Big Bang theory postulates that the entire presently known universe, shortly after the Big Bang, was approximately 1 AU in radius (1.5×10^{11} m) and had a mass density of 10^{15} g/cm^3. If we assume that one part of this sphere was composed of protons, one part of neutrons, and one part of electrons, how many particles were contained in the universe at that time? (Protons and neutrons each have a mass of 1.67×10^{-30} g. An electron has a mass of 9.11×10^{-33} g.)

500. A nuclear power plant generates 3,000 MW of power daily. If the fission of one uranium^{235} atom releases 200 MeV, how many fissions take place each second? How much mass is converted to energy daily? (Neglect friction in the system.)

ANSWERS

Chapter 1: Vectors

1. (C) ($13.\hat{\mathbf{i}} - 6.\hat{\mathbf{j}}$) in which all the $\hat{\mathbf{i}}$ components are added together to get the resultant vector $\hat{\mathbf{i}}$ component and similarly with the $\hat{\mathbf{j}}$ component.

2. (C) They state a direction but not a magnitude.

3. (D) It states a magnitude but not a direction. Although it is one dimensional and has a direction of change, temperature is only a magnitude and, therefore, a scalar.

4. (D) Speed states only a magnitude, and velocity states a magnitude and direction. Although (B) is a true statement, it is not an explanation of the difference between the two terms.

5. (C) This is because magnitude is expressed by R, and direction is expressed by θ and ϕ.

6. (A) 1,520 is the correct answer when considering significant figures. Vector magnitude is the square root of the sum of the squared components.

7. (A) 9° is the correct answer when considering significant figures. The angle is given by taking the arctangent of the y component divided by the x component.

8. (C) $|A| \cdot |B| \cdot \cos\theta$.

9. (C) $|A| \cdot |B| \cdot \sin\theta$ directed perpendicular to the plane of **A** and **B**. Because **A** is first in the equation, using the right-hand rule and sweeping **A** into **B** gives the direction from the plane containing **A** and **B**.

10. (D) Moles.

11. (D) 42 and 27 account for significant figures. x component = magnitude × cos (33°) and y component = magnitude × sin (33°).

12. (D) $42.\hat{\mathbf{i}} + 27.\hat{\mathbf{j}} - 29.\mathbf{k}$. Using the magnitude, first calculate the z component and the component in the x–y plane. Then using the x–y plane and the x component, calculate the x and y components.

13. (B) $41.\hat{\mathbf{i}} + 10.\hat{\mathbf{j}} + 98.\mathbf{k}$ is obtained by adding the individual x, y, and z components.

14. (C) 0.1 mi $\hat{\mathbf{i}}$ + 1.1 mi $\hat{\mathbf{j}}$. While the hiker is 1.1 miles from his starting point, a vector quantity was requested.

15. (C) Force. All the others listed are scalar quantities.

16. (D) (A), (B), and (C) are just three of the many uses of vectors.

17. (E) 38.4°. First, the magnitude of the *x–y* plane component must be calculated; it is 67.6. Then, take the arctangent of the *x* component divided by the *x–y* plane magnitude. The result is arcos (53/67.6) = 38.4°.

18. (B) 66.8°. Similar to **Question 17**, the magnitude of the vector must be calculated, which is 73.6. Then, take the arcos (29/73.6) = 66.8°.

19. (B) A vector.

20. (D) A scalar or a vector depending on whether the multiplication is the dot product or the cross product.

21. (B) –546.0. In this case, the dot product is the product of the two magnitudes.

22. (C) 0.0. In this case, the two vectors have a combined angle of 180°. Then, the magnitude of the cross product vector is $|A| \cdot |B| \cdot \sin\theta = 0$, with $\theta = 0°$.

23. (D) It defines the positive or negative direction of the cross product vector.

24. (C) 0.0. The dot product is $|A| \cdot |B| \cdot \cos\theta = 0$, with $\theta = 90°$.

25. (B) 546.0 **k**. The cross product vector is $|A| \cdot |B| \cdot \sin\theta = 546$, with $\theta = 90°$, and a direction in the positive *z* direction.

26. (D) No, because the only information provided is a direction without a magnitude.

27. (B) A + B = B + A.

28. (D) The cross product gives a vector perpendicular to A and a cos 90° = 0.

29. $\mathbf{A} = 13\ \text{N}\,\hat{\imath} + 11\ \text{N}\,\hat{\jmath} + 17\ \text{N}\,\text{k}$ and $\mathbf{B} = -7.5\ \text{N}\,\hat{\imath} + 13\ \text{N}\,\hat{\jmath} + 15\ \text{N}\,\text{k}$. $A \cdot B = 300.\ \text{N}^2$. $\mathbf{A} \times \mathbf{B} = -56\ \text{N}\,\hat{\imath} - 322\ \text{N}\,\hat{\jmath} + 252\ \text{N}\,\text{k}$.

30. Using Earth and Alkaid as a reference direction, Merak is 25.6° from Alkaid. So, with that information we know two vector magnitudes and the angle between those vectors. Adding the two vectors gives us the distance from Alkaid to Merak. Then, 138 ly cos (25.6°) gives us the vector component along the reference line from Earth to Alkaid, and 138 ly sin (25.6°) gives us the vector component perpendicular to the reference line. Then, find the magnitude by adding the vector components and taking the square root of the squares of the components. This result is 210 light years between Alkaid and Merak.

Chapter 2: Free-Body Diagrams and Equilibrium

31. (A) The first, which mathematically states that $\Sigma \mathbf{F} = 0$.

32. (D) According to Newton's third law, it is the weight of the book and the pushback force of the table.

33. (C) When the object is on an inclined plane. The normal force is then a function of the slope of the inclined plane and will be less than the object's weight. In the other cases, the object either has an equal or nonexistent normal force.

34. (C) A free-body diagram (FBD) of the string shows that an angle must exist between the string and a perfectly straight line to allow vertical forces to balance the weight of the string.

35. (E) $\Sigma F_x = 0$, $\Sigma F_y = 0$, $\Sigma F_z = 0$, $\Sigma M_x = 0$, $\Sigma M_y = 0$, and $\Sigma M_z = 0$ is the complete expression of Newton's first law for the x, y, z coordinate system.

36. (C) A force applied to a body that causes the body to rotate. Some of the other answers do not completely explain the definition of torque.

37. (E) Generally, no. A rope can carry a compressive force if that force is applied perpendicular to the rope, but that is an extremely rare case.

38. (C) It is a force that is proportional to the normal force that acts in opposition to the direction of motion.

39. (C) An FBD shows the forces and accelerations and their locations acting on a body.

40. (D) The distance perpendicular to the line of action of the force from the pivot point.

41. (B) 200 pounds. The pulley simply redirects the force applied to the rope by the box.

42. (B) The answer is 498 N when using the proper significant figures. The calculated answer is 497.5, but it needs to be rounded to 498 to be correct.

43. (D) $T_{right} = 1{,}610$ N and $T_{left} = 2{,}630$ N. There are two unknowns (i.e., the tensions in each rope), but there are two equations that define the x and y components of the tensions. Solving both equations will yield the above results when using significant figures.

44. (D) 1,000 N. Using the summation of moments about the tip yields the force applied 2 cm from the tip.

45. (C) 26° is found from the arctan, (0.5) = 26°.

46. (E) The astronaut moves in the direction of the force on the wrench. With no gravity to restrain him, the astronaut will move in response to the forces applied to him or those generated by him. In this case, the bolt cannot be loosened unless the astronaut is somehow fixed so he can apply force to the wrench. Otherwise, the force he applies will cause him to move in the direction of the force. The massive satellite is unlikely to rotate because of its high moment of inertia.

47. (B) 100 N is found by the summation of moments about the rear wheels. Account for significant figures.

48. (C) 1,633 N. A six-strand pulley has six supporting cables. Thus, the tension in the cable, which is continuous through the pulleys, is simply the total load divided by 6.

49. (B) *Philosophiæ Naturalis Principia Mathematica*, which was originally published in 1687.

50. (D) The weight of the object in the downward direction and the supporting force of the spring in the upward direction.

51. (E) The weight of the box, the resultant force of the plane on the box, and the friction force along the surface of the plane beneath the box. These are the only forces acting on the box. The others mentioned are components or they are misleading.

52. (A) 213 N is found by the summation of moments about the lift point.

53. (C) 1,760 N and 1,960 N. In (A) and (B), the buoyancy force caused by displacing the water acts opposite to the weight of the chest, so the force to pull up the chest is 1,760 N. Once out of the water, the full weight of the chest must be supported, so a force of 1,960 N is needed.

54. (C) 333 N is found by the summation of moments about the fulcrum. Since a vertical force is called for, accounting for the angle is unnecessary.

55. (D) 4.73×10^6 m from the Earth. This problem is similar to the question about the summation of moments, except that the masses are used instead of the weights.

56. (B) 682 N. The pulley effectively divides the weight of the block along the plane by 2.

57. (D) At the front center of the machine. That way, all three men are pushing the machine straight into the niche.

58. (B) 50 N. This problem depends on the given distances, which form a triangle with an angle of 0.5723° between the vertical rope position and the position where the bridge section is to be moved. The required force is the sin of that angle, then multiplied by the weight of the bridge section.

59. 1,048 N is the tension in the rope. This includes the weight along the plane, the friction force from the normal weight to the plane, and the pulley dividing the force by 2. This problem is similar to **Question 26**, but this question includes friction.

60. This is a simple summation of moments problem, but it is slightly obscured by the pulley and the use of significant figures. The answer is less than 26,000 N.

Chapter 3: Kinematics

61. (B) You know the height (120 m) and the initial velocity (0 m/s). The acceleration due to gravity (9.8 m/s²) is implied. Solve this equation for t:

$$\Delta x = v_0 t + \frac{1}{2}at^2$$

$$(125\ \text{m}) = (0\ \text{m/s}) + \frac{1}{2}(9.8\ \text{m/s}^2)t^2$$

$$t = 5.0\ \text{s}$$

62. (D) You know the distance the car traveled (500 m), the initial velocity (0 m/s), and the final velocity (50 m/s). Solve this equation for a:

$$v_f^2 = v_0^2 + 2a\Delta x$$

$$(50\ \text{m/s})^2 = (0\ \text{m/s})^2 + 2a(500\ \text{m})$$

$$a = 2.5\ \text{m/s}^2$$

63. (C) The horizontal velocity of the package is the same as that of the plane (50.0 m/s). The height of the plane is 125.0 m. You must first find the time it takes for the package to reach the ground and use that time to calculate the horizontal distance that the package travels.

1. Solve for time:

   $$\Delta y = y_0 t - \frac{1}{2}gt^2$$

   $$(-125\ \text{m}) = (0\ \text{m/s}) - \frac{1}{2}(9.8\ \text{m/s}^2)t^2$$

   $$t = 5\ \text{s}$$

2. Solve for horizontal distance:

   $$\Delta x = v_x t$$

   $$\Delta x = (50.0\ \text{m/s})(5\ \text{s})$$

   $$\Delta x = 250\ \text{m}$$

64. (D) You know the angle of the kick (30°) and the initial velocity (10.0 m/s). You must find the vertical component of the velocity and solve for t when the ball hits the ground ($y = 0$).

1. Solve for vertical component of velocity:

$$v_y = v \sin\theta$$

$$v_y = (10.0 \text{ m/s})(\sin 30°)$$

$$v_y = 5.0 \text{ m/s}$$

2. Solve for time:

$$\Delta y = v_y t + \frac{1}{2} g t^2$$

$$0 = v_y t + \frac{1}{2} g t^2$$

$$0 = t\left(v_y + \frac{1}{2} g t\right) : \ t = 0 \text{ is not a relevant solution}$$

$$v_y + \frac{1}{2} g t = 0$$

$$t = \frac{-2v_y}{g}$$

$$t = \frac{-2(5.0 \text{ m/s})}{(-9.8 \text{ m/s}^2)}$$

$$t = 1.0 \text{ s}$$

65. (C) The object is at rest during the time interval of 3–4 s when the object's position does not change.

66. (D) You know the car's final velocity (30 m/s), the rate of acceleration (3.0 m/s²), and the time interval over which it accelerated (5.0 s). You can calculate the initial velocity:

$$v_f = v_0 + at$$

$$v_0 = v_f - at$$

$$v_0 = (30 \text{ m/s}) - (3.0 \text{ m/s}^2)(5 \text{ s})$$

$$v_0 = 15 \text{ m/s}$$

67. (C) The airplane's initial velocity was 0 m/s and it accelerated at 1.0 m/s² for 5 min (300 s). You can calculate the distance:

$$\Delta x = v_0 t + \frac{1}{2} a t^2$$

$$\Delta x = 0 + \frac{1}{2}(1.0 \text{ m/s}^2)(300 \text{ s})^2$$

$$\Delta x = 45{,}000 \text{ m or } 45 \text{ km}$$

68. (B) By subtracting the echo time, you can calculate that the stone takes 8 s to reach the bottom of the well. Now, you can calculate the vertical distance of the well:

$$\Delta x = v_0 t + \frac{1}{2}at^2$$

$$\Delta x = 0 + \frac{1}{2}(9.8 \text{ m/s}^2)(8 \text{ s})^2$$

$$\Delta x = 320 \text{ m}$$

69. (E) At the maximum height, the velocity (v_f) is zero. So, solve the equation $v_f = v_0 +$ at for t.

70. (B) To get the velocity, take the derivative of $x(t) = 4t^2 + 6t + 2$ and you will get $v(t) = 8t + 6$. At $t = 6$ s, $v = 8(t) + 6 = 38$ m/s. To find the original direction of motion, evaluate the expression for $t = 0$ s. You get $v = 8(0) + 6 = 6$ m/s.

71. (E) The initial velocity is 0 m/s, the height is 15 m, and the time is 1 s. You can calculate the acceleration due to gravity:

$$\Delta x = v_0 t + \frac{1}{2}at^2$$

$$15 \text{ m} = 0 + \frac{1}{2}(a)(1 \text{ s})^2$$

$$a = 30 \text{ m/s}^2$$

72. (A) You know the car's initial velocity (0 m/s), the rate of acceleration (3.0 m/s^2), and the distance of the track (150 m). You can calculate the final velocity:

$$v_f^2 = v_0^2 + 2a\Delta x$$

$$v_f^2 = (0 \text{ m/s})^2 + 2(3.0 \text{ m/s}^2)(150 \text{ m})$$

$$v_f = 30 \text{ m/s}$$

73. (D) The first derivative of a position–time function is a velocity–time function. The velocity at any instant in time is the slope of the line tangent to any point on the position–time graph. When the tangent line is horizontal (slope = 0), then the velocity is zero at that point. For this graph, that occurs at 1 s and 3 s.

74. (C) A projectile launched at an angle from the ground follows a parabolic path. The parabola is symmetrical, so the maximum height occurs at one-half of the total time in the air.

75. **(D)** It takes approximately 3 s for the stone to drop and the stone started from rest. So, we can calculate the approximate height of the cliff:

$$\Delta x = v_0 t + \frac{1}{2}at^2$$

$$\Delta x = 0 + \frac{1}{2}(9.8 \text{ m/s}^2)(3 \text{ s})^2$$

$$\Delta x = 45 \text{ m}$$

76. **(D)** The bicycle's initial velocity was 5.0 m/s, and it slowed at an acceleration of -2.5 m/s^2 to a stop ($v_f = 0$). The time can be determined from this equation:

$$v_f = v_0 + at$$

$$(0 \text{ m/s}) = (5.0 \text{ m/s}) + (-2.5 \text{ m/s}^2)t$$

$$t = 2.0 \text{ s}$$

77. **(B)** The final velocity of the car was zero and the rate of deceleration was −7.5 m/s^2 across a distance of 60 m. You can calculate the initial velocity of the car when it started skidding:

$$v_f^2 = v_0^2 + 2a\Delta x$$

$$(0 \text{ m/s})^2 = v_0^2 + 2(-7.5 \text{ m/s}^2)(60 \text{ m})$$

$$v_0 = 30 \text{ m/s}$$

78. **(A)** To find the displacement, you must integrate the area under the curve to the t axis. This simply involves adding the areas of two trapezoids, one in the 0–6 s time interval and the other in the 6–10 s time interval.

$$\Delta x = A_{0-6s} + A_{6-10s}$$

$$\Delta x = \left[\frac{1}{2}(b_1 + b_2)(h)\right] + \left[\frac{1}{2}(b_1 + b_2)(h)\right]$$

$$\Delta x = \left[\frac{1}{2}(3 \text{ s} + 6 \text{ s})(2 \text{ m/s})\right] + \left[\frac{1}{2}(2 \text{ s} + 4 \text{ s})(-2 \text{ m/s})\right]$$

$$\Delta x = [9 \text{ m}] + [-6 \text{ m}]$$

$$\Delta x = 3 \text{ m}$$

79. **(B)** The horizontal component of velocity is calculated by $v_x = v \cos \theta$, while the vertical component is calculated by $v_x = v \sin \theta$. For angles less than 45°, the cosine is larger than the sine. So, the horizontal component of velocity is larger than the vertical component.

80. (B) The dropped ball's initial velocity was zero and it took 4 s to hit the ground. The height of the building can be calculated:

$$\Delta x = v_0 t + \frac{1}{2}at^2$$

$$\Delta x = 0 + \frac{1}{2}(9.8 \text{ m/s}^2)(4 \text{ s})^2$$

$$\Delta x = 90 \text{ m}$$

81. (D) If the positive direction is forward, then an object that is moving backward at a constant velocity would have a position–time graph that is linear with a negative slope.

82. (C) For projectiles launched at the same velocity, but at different angles (between 0° and 90°), those with complementary angles will have the same range.

83. (D) You know the initial velocity (0 m/s), the acceleration (20 m/s^2), and the time (2 min = 120 s). Therefore:

$$v_f = v_0 + at$$

$$v_f = (0 \text{ m/s}) + (20 \text{ m/s}^2)(120 \text{ s})$$

$$v_f = 2400 \text{ m/s}$$

84. (B) The initial velocity of the car is 30 m/s, the final velocity is 0 m/s, and the acceleration is –9 m/s^2. The distance the car traveled can be calculated:

$$v_f^2 = v_0^2 + 2a\Delta x$$

$$(0 \text{ m/s})^2 = (30 \text{ m/s})^2 + 2(-9 \text{ m/s}^2)\Delta x$$

$$\Delta x = 50 \text{ m}$$

85. (C) The position in the graph is increasing nonlinearly in the positive direction. It is consistent with motion with a constant positive acceleration.

86. (B) The initial velocity of the ball is 0 m/s and the time is 2 s. The distance it travels can be calculated:

$$\Delta x = v_0 t + \frac{1}{2}at^2$$

$$\Delta x = 0 + \frac{1}{2}(9.8 \text{ m/s}^2)(2 \text{ s})^2$$

$$\Delta x = 20 \text{ m}$$

87. (D) The ice skater has an initial velocity of 10 m/s and a final velocity of 0 m/s. The time interval is 0.5 s. Calculate the rate of acceleration:

$$v_f = v_0 + at$$

$$(0 \text{ m/s}) = (10 \text{ m/s}) + a(0.5 \text{ s})$$

$$a = -20 \text{ m/s}^2$$

88. (D) The position in the graph is increasing linearly in the positive direction. It is consistent with motion with a constant positive velocity and, hence, zero acceleration.

89. This is a projectile motion problem where $v = 10.0$ m/s, $\Delta y = -45$ m, and $\theta = 45°$.

(a) The arrow follows a parabolic path. It first moves upward and away from the wall. It peaks and then moves downward and away from the wall until it hits the ground.

(b) The horizontal component of velocity is $v_x = v \cos\theta = (10.0 \text{ m/s}) \cos 45° = 7.1$ m/s. The vertical component of velocity is $v_y = v \sin\theta = (10.0 \text{ m/s}) \sin 45° = 7.1$ m/s. Because the cosine and sine of 45° are equal, the horizontal and vertical components are equal.

(c) The time it takes for the arrow to reach the ground can be calculated as follows:

$$\Delta y = v_y t + \frac{1}{2} g t^2$$

$$-45 \text{ m} = (7.1 \text{ m/s})t + \frac{1}{2}(-9.8 \text{ m/s}^2)t^2$$

When you solve the quadratic equation by the quadratic formula, you get

$$t = 3.84 \text{ s}$$

(d) The range of the arrow can be calculated as follows:

$$\Delta x = v_x t$$

$$\Delta x = (7.1 \text{ m/s})(3.84 \text{ s})$$

$$\Delta x = 27.3 \text{ m}$$

90. Starting with this graph:

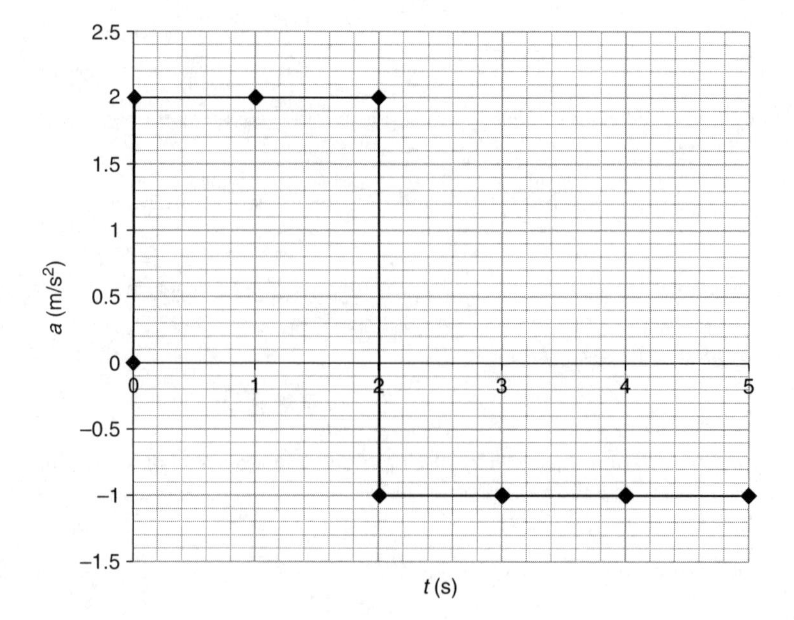

(a) Velocity is calculated by integrating the area under the curve of the acceleration–time graph (i.e., between the curve and the x axis).

t (s)	v (m/s)
0	0
1	2
2	4
3	3
4	2
5	1

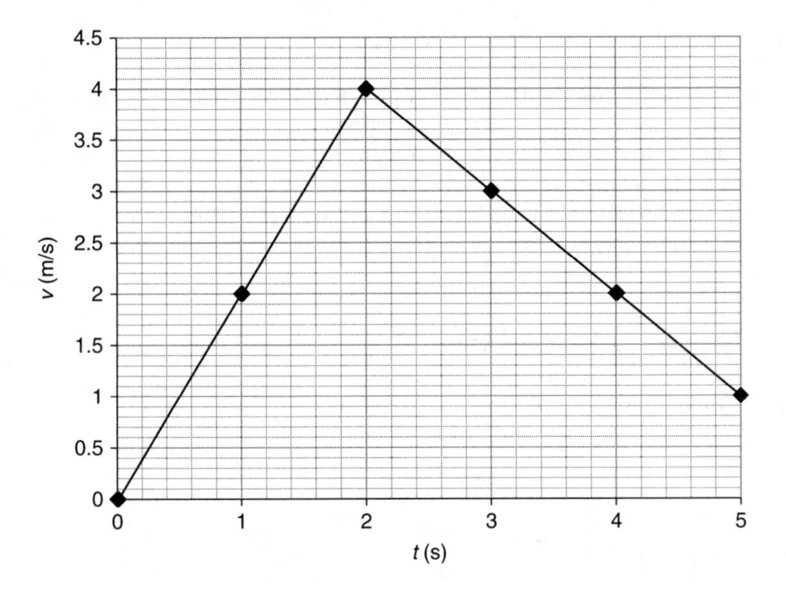

(b) The displacement intervals are calculated by integrating the area under the velocity–time graph. The position–time graph is plotted by adding the displacements.

t (s)	Δx (m)	x (m)
0	0	0
1	1	1
2	3	4
3	3.5	7.5
4	2.5	10
5	1.5	11.5

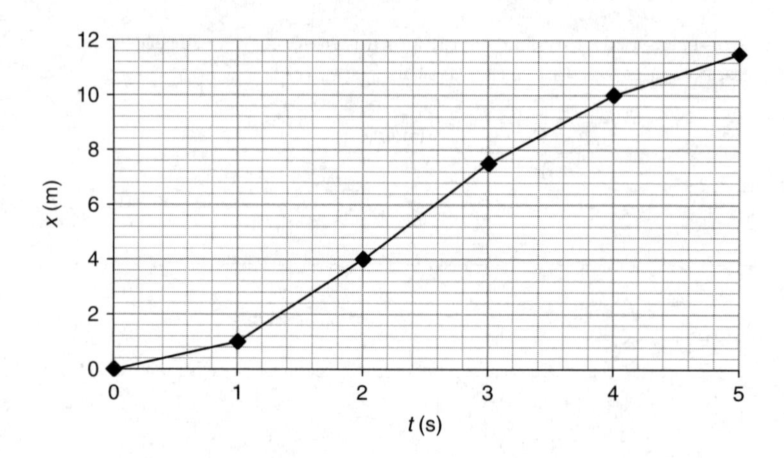

(c) The bicycle begins from rest at the starting point. It accelerates constantly for 2 s and the velocity increases linearly. The bicycle travels away from the starting point and the distance increases exponentially. From 2 to 5 s, the bicycle's rate of acceleration decreases to a constant, but negative value. The bicycle's velocity decreases. The bicycle continues to travel away from the starting line, but the distance gets smaller with time.

Chapter 4: Newton's Second Law

91. (B) A 10.0-N force pushes the box to the right, while an 8.0-N force pushes it to the left. The mass of the box is 5 kg. Solve this equation for *a*:

$$F_{net} = ma$$

$$(10.0 \text{ N} - 8.0 \text{ N}) = (5 \text{ kg})a$$

$$a = 0.4 \text{ m/s}^2 \text{ to the right}$$

92. (D) Remember that the force of gravity (*mg*) must be resolved into *x* and *y* components, and that the force of friction (F_f) is related to the normal force (F_N) by $F_f = \mu F_N$. The box does not move above or into the plane, so the forces in the *y* direction must be balanced. You can then apply Newton's second law to both the *x* and *y* directions to come up with an expression for acceleration:

$$F_{nety} = 0$$

$$F_N = mg \cos \theta = 0$$

$$F_N = mg \cos \theta$$

$$F_{netx} = ma$$
$$mg \sin\theta - F_f = ma$$
$$mg \sin\theta - \mu F_N = ma$$
$$mg \sin\theta - \mu(mg \cos\theta) = ma$$
$$g \sin\theta - \mu(g \cos\theta) = a$$
$$a = g(\sin\theta - \mu \cos\theta)$$

93. (E) The force that you exert on the rope is equal to the rope's tension (T). The acceleration due to gravity is approximately 10 m/s^2. The acceleration of the box is given at 10 m/s^2 upward. Use Newton's second law to calculate the downward force that you must apply to the rope:

$$F_{net} = ma$$
$$T - mg = ma$$
$$T = ma + mg$$
$$T = m(a + g)$$
$$T = (10 \text{ kg})(10 \text{ m/s}^2 + 10 \text{ m/s}^2)$$
$$T = 200 \text{ N}$$

94. (A) You must resolve the jet's thrust into horizontal and vertical components. Then enter the horizontal component of thrust, force of the wind, and the jet's mass into Newton's second law to find the acceleration:

$$F_x = F \cos\theta$$
$$F_x = (20{,}000 \text{ N})(\cos 60°)$$
$$F_x = 10{,}000 \text{ N}$$
$$F_{netx} = ma$$
$$a = \frac{F_{netx}}{m}$$
$$a = \frac{(F_x - F_{wind})}{m}$$
$$a = \frac{(10{,}000 \text{ N} - 1000 \text{ N})}{(90{,}000 \text{ kg})}$$
$$a = 0.1 \text{ m/s}^2$$

95. (C) The object accelerates only when the forces acting upon it are unbalanced. Of the choices of times listed, only 4 s and 8 s both have unbalanced forces.

96. (C) Acceleration is the second derivative of the position function. According to Newton's second law, when acceleration is zero, there is no net force. So, calculate the acceleration function, set it to zero, and solve it for *t*:

Position function: $x(t) = \frac{1}{3}t^3 - 2t^2 + 3t + 2$

First derivative: $v(t) = t^2 - 4t + 3$

Second derivative: $a(t) = 2t - 4$

Set second derivative to zero and solve for t:

$0 = 2t - 4$

$2t = 4$

$t = 2\text{ s}$

97. (D) The crate's mass is 10 kg, one force is 10 N, and the rate of acceleration is 0.1 m/s^2. You can solve for the second force by using Newton's second law:

$$a = \frac{F_{net}}{m}$$

$$a = \frac{(F_1 - F_2)}{m}$$

$$F_1 - F_2 = ma$$

$$F_2 = F_1 - ma$$

$$F_2 = (10\text{ N}) - (10\text{ kg})(0.1\text{ m/s}^2)$$

$$F_2 = 9\text{ N}$$

98. (C) By applying Newton's second law to the inclined situation, you can derive an equation in which the crate's acceleration depends only upon the coefficient of friction ($\mu = 0.1$), acceleration due to gravity ($g \approx 10$ m/s^2), and the angle of incline ($\theta = 45°$).

$$F_{nety} = 0$$

$$F_N - mg\cos\theta = 0$$

$$F_N = mg\cos\theta$$

$$F_{netx} = ma$$

$$mg\sin\theta - F_f = ma$$

$$mg\sin\theta - \mu F_N = ma$$

$$mg\sin\theta - \mu(mg\cos\theta) = ma$$

$$g\sin\theta - \mu(g\cos\theta) = a$$

$$a = g(\sin\theta - \mu\cos\theta)$$

$$a = (10\text{ m/s}^2)[\sin 45° - (0.1)\cos 45°]$$

$$a = (10\text{ m/s}^2)[(0.707) - (0.1)(0.707)]$$

$$a = 6.4\text{ m/s}^2$$

99. (B) The two masses are designated $m = 5$ kg and $M = 10$ kg. By applying Newton's second law to the pulley system and using the direction of positive as counterclockwise, you can derive an equation for acceleration and solve the problem:

$$mg - T = ma$$
$$T - Mg = Ma$$
$$T = Ma + Mg$$
$$mg - (Ma + Mg) = ma$$
$$mg - Ma - Mg = ma$$
$$mg - Mg = ma + Ma$$
$$g(m - M) = a(m + M)$$
$$a = g\frac{(m - M)}{(m + M)}$$
$$a = (10\ \text{m/s}^2)\frac{(5\ \text{kg} - 10\ \text{kg})}{(5\ \text{kg} + 10\ \text{kg})}$$
$$a = -3.3\ \text{m/s}^2$$

100. (A) From the graph, the acceleration at $t = 1$ s is -2 m/s^2. The object's mass is 5 kg. So, the net force can be calculated from Newton's second law:

$$F_{net} = ma$$
$$F_{net} = (5\ \text{kg})(-2\ \text{m/s}^2)$$
$$F_{net} = -10\ \text{N}$$

101. (C) You must find the x component of the skier's weight. The skier has a mass of 70 kg, the angle of the incline is 30°, and the skier pushes off with an applied force ($F_A = 105$ N). Use Newton's second law to find the rate of acceleration:

$$F_{netx} = ma$$
$$mg\sin\theta + F_A = ma$$
$$a = \frac{mg\sin\theta + F_A}{m}$$
$$a = \frac{(70\ \text{kg})(10\ \text{m/s}^2)\sin 30° + (105\ \text{N})}{(70\ \text{kg})}$$
$$a = \frac{(70\ \text{kg})(10\ \text{m/s}^2)(0.5) + (105\ \text{N})}{(70\ \text{kg})}$$
$$a = 2\ \text{m/s}^2$$

102. (B) You know the bullet's initial velocity (0 m/s) and mass (0.010 kg). You must use Newton's second law to calculate the rate of the bullet's acceleration. The bullet uniformly accelerates over the distance of the musket barrel (1 m) to its final velocity:

$$a = \frac{F_{net}}{m}$$

$$a = \frac{(50 \text{ N})}{(0.010 \text{ kg})}$$

$$a = 5{,}000 \text{ m/s}^2$$

$$v_f^2 = v_0^2 + 2a\Delta x$$

$$v_f^2 = (0 \text{ m/s})^2 + 2(5{,}000 \text{ m/s}^2)(1 \text{ m})$$

$$v_f = 100 \text{ m/s}$$

103. (C) The second derivative of a position–time function is an acceleration–time function. When the acceleration is zero, there is no net force acting on the object. From a position–time graph, the inflection point (where the curve changes the direction of concavity) is where the acceleration is zero. For this graph that occurs at 2 s.

104. (C) The two masses are designated $m = 10$ kg and $M = 30$ kg. By applying Newton's second law to the pulley system and using the direction of positive as counterclockwise, you can derive an equation for acceleration and solve the problem:

$$F_{netx} = 0 - T = Ma\text{: Block } M$$

$$F_{net} = T - mg = ma\text{: Block } m$$

$$(-Ma) - mg = ma$$

$$ma = -Ma - mg$$

$$ma + Ma = -mg$$

$$a(m + M) = -mg$$

$$a = \frac{-mg}{(m + M)}$$

$$a = \frac{-(10 \text{ kg})(10 \text{ m/s}^2)}{(10 \text{ kg} + 30 \text{ kg})}$$

$$a = -2.5 \text{ m/s}^2$$

105. (C) The jet has a mass of 50,000 kg, the engines produce 20,000 N of thrust, and the acceleration is 0.3 m/s^2. From Newton's second law, you can calculate the force of the air resistance:

$$F_{net} = ma$$

$$F_{\text{thrust}} - F_{\text{air resistance}} = ma$$

$$-F_{\text{air resistance}} = ma - F_{\text{thrust}}$$

$$F_{\text{air resistance}} = F_{\text{thrust}} - ma$$

$$F_{\text{air resistance}} = (20{,}000 \text{ N}) - (50{,}000 \text{ kg})(0.3 \text{ m/s}^2)$$

$$F_{\text{air resistance}} = 5{,}000 \text{ N}$$

106. (E) The acceleration is the slope of a velocity–time graph. In this case, the slope of the graph yields an acceleration of 0.5 m/s^2. The mass of the crate is 50 kg. From Newton's second law, you can calculate the force applied:

$$F_{net} = ma$$

$$F_{net} = (50 \text{ kg})(0.5 \text{ m/s}^2)$$

$$F_{net} = 25 \text{ N}$$

107. (E) You know the initial velocity of the car (30 m/s), the final velocity of the car (0 m/s), and the distance over which the car uniformly decelerates ($\Delta x = 60$ m). You can calculate the car's acceleration. Once you know the acceleration and the car's mass (1,000 kg), you can use Newton's second law to calculate the net force on the car.

$$v_f^2 = v_0^2 + 2a\Delta x$$

$$0 = v_0^2 + 2a\Delta x$$

$$-2a\Delta x = v_0^2$$

$$a = -\frac{v_0^2}{2\Delta x}$$

$$a = -\frac{(30 \text{ m/s})^2}{2(60 \text{ m})}$$

$$a = -7.5 \text{ m/s}^2$$

$$F_{net} = ma$$

$$F_{net} = (1{,}000 \text{ kg})(-7.5 \text{ m/s}^2)$$

$$F_{net} = -7{,}500 \text{ N}$$

108. (D) You must first calculate the acceleration using Newton's second law. The net force is 100 N and the mass of the ball is 5 kg. Next, you must calculate the final velocity when the ball starts from rest ($v_0 = 0$ m/s) and the time interval is 1.5 s:

$$a = \frac{F_{net}}{m}$$

$$a = \frac{(100 \text{ N})}{(5 \text{ kg})}$$

$$a = 20 \text{ m/s}^2$$

$$v_f = v_0 + at$$

$$v_f = (0 \text{ m/s}) + (20 \text{ m/s}^2)(1.5 \text{ s})$$

$$v_f = 30 \text{ m/s}$$

109. (B) The force of the opposing wind will cause the football to decelerate horizontally. So, the horizontal velocity will decrease. This would not be the case if there was no wind. Because the wind acts horizontally, the vertical components of the football's path (e.g., velocity, height) will not be affected.

110. (B) The opposing forces acting upon the skydiver are gravity and air resistance. Initially, the net force is high (gravity > air resistance), the skydiver accelerates, and the velocity increases rapidly. Over time, the air resistance increases and the net force decreases. So, the acceleration decreases. At some point in time, there is no net force (gravity = air resistance), acceleration drops to zero, and the skydiver's velocity is constant (e.g., terminal velocity).

111. (D) You know the box's initial velocity (0 m/s) and mass (10 kg). You also know her applied force (50 N) and the force of friction opposing her (45 N). You must use Newton's second law to calculate the rate of the box's acceleration. The box uniformly accelerates to its final velocity of (2 m/s). You can calculate the time:

$$a = \frac{F_{net}}{m}$$

$$a = \frac{F_A - F_f}{m}$$

$$a = \frac{(50\text{ N} - 45\text{ N})}{(10\text{ kg})}$$

$$a = 0.5\text{ m/s}^2$$

$$v_f = v_0 + at$$

$$2.0\text{ m/s}^2 = 0\text{ m/s} + (0.5\text{ m/s}^2)t$$

$$t = 4\text{ s}$$

112. (E) The rocket goes from rest ($v_0 = 0$ m/s) to 9.6 km/s ($v_f = 9600$ m/s) in 8 min ($t = 480$ s). The rocket's mass is 8.0×10^6 kg. First, determine the rate of acceleration and then use Newton's second law to determine the net force:

$$v_f = v_0 + at$$

$$9{,}600\text{ m/s} = 0\text{ m/s} + a(480\text{ s}^2)t$$

$$a = 20\text{ m/s}^2$$

$$F_{net} = ma$$

$$F_{net} = (8.0 \times 10^6\text{ kg})(20\text{ m/s}^2)$$

$$F_{net} = 9.6 \times 10^7\text{ N}$$

113. (A) The car travels at constant velocity (its value is irrelevant). So, the net force acting on the car is zero. Therefore, the force of friction is equal to the thrust of the engine (1,000 N). The mass of the car is 1,000 kg. So calculate the coefficient of friction:

$$F_f = \mu F_N$$

$$F_{nety} = F_N - mg$$

$$0 = F_N - mg$$

$$F_N = mg$$

$$F_f = \mu mg$$

$$\mu = \frac{F_f}{mg}$$

$$\mu = \frac{(1{,}000 \text{ N})}{(1{,}000 \text{ kg})(10 \text{ m/s}^2)}$$

$$\mu = 0.1$$

114. (B) The initial velocity of the car is 30 m/s, the final velocity is 0 m/s, and the net force acting on the car is 9,000 N. The car's mass is 1,000 kg. Use Newton's second law to calculate the rate of acceleration (negative as implied by deceleration). The distance the car traveled can be calculated:

$$a = \frac{F_{net}}{m}$$

$$a = \frac{(-9{,}000 \text{ N})}{(1{,}000 \text{ kg})}$$

$$a = -9 \text{ m/s}^2$$

$$v_f^2 = v_0^2 + 2a\Delta x$$

$$(0 \text{ m/s})^2 = (30 \text{ m/s})^2 + 2(-9 \text{ m/s}^2)\Delta x$$

$$\Delta x = 50 \text{ m}$$

115. (D) The position in the graph is increasing linearly in the positive direction. It is consistent with motion with a constant positive velocity. The constant velocity means that acceleration is zero. According to Newton's second law, when acceleration is zero, there is no net force acting upon it.

116. (D) The initial velocity of the runner is 0 m/s, the final velocity is 10 m/s, and the time is 0.5 s. The runner's mass is 70 kg. First, calculate the acceleration and use Newton's second law to find the net force:

$$v_f = v_0 + at$$

$$(10 \text{ m/s}) = (0 \text{ m/s}) + a(0.5 \text{ s})$$

$$a = 20 \text{ m/s}^2$$

$$F_{net} = ma$$

$$F_{net} = (70 \text{ kg})(20 \text{ m/s}^2)$$

$$F_{net} = 1{,}400 \text{ N}$$

117. (B) The ice skater has a mass of 50 kg. The force of friction acts opposite of her direction of motion ($F_f = -1{,}000$ N). Her deceleration is uniform to a final velocity of 0 m/s in 0.5 s, so you can calculate her acceleration from Newton's second law and then find her initial velocity:

$$a = \frac{F_{net}}{m}$$

$$a = \frac{F_f}{m}$$

$$a = \frac{(-1{,}000 \text{ N})}{(50 \text{ kg})}$$

$$a = -20 \text{ m/s}^2$$

$$v_f = v_0 + at$$

$$0 \text{ m/s}^2 = v_0 + (-20 \text{ m/s}^2)(0.5)$$

$$v_0 = 10 \text{ m/s}$$

118. (C) According to Newton's second law, the forces acting upon an object are balanced when the acceleration is zero. As shown in the graph, this occurs at 2 s.

119. This is a projectile motion problem where $v = 10.0$ m/s, $\Delta y = -45$ m, and $\theta = 45°$. The arrow ($m = 0.05$ kg) travels against a wind in the x direction with a force of -0.05 N.

(a) The horizontal component of velocity is $v_x = v \cos\theta = (10.0 \text{ m/s}) \cos 45° = 7.1$ m/s. The vertical component of velocity is $v_y = v \sin\theta = (10.0 \text{ m/s}) \sin 45° = 7.1$ m/s. Because the cosine and sine of 45° are equal, the horizontal and vertical components are equal.

(b) The time it takes for the arrow to reach the ground can be calculated as follows:

$$\Delta y = v_y t + \frac{1}{2} g t^2$$

$$-45 \text{ m} = (7.1 \text{ m/s})t + \frac{1}{2}(-9.8 \text{ m/s}^2)t^2$$

When you solve the quadratic equation by the quadratic formula, you get:

$$t = 3.84\text{s}$$

(c) While in flight, the wind exerts a constant net force of -0.05 N against the arrow. This will cause the arrow to slow down in the horizontal direction. You can calculate the arrow's acceleration using Newton's second law:

$$a = \frac{F_{net}}{m}$$

$$a = \frac{-0.05 \text{ N}}{-0.05 \text{ kg}}$$

$$a = -1.0 \text{ m/s}^2$$

You can calculate the final velocity of the arrow:

$$v_f = v_0 + at$$

$$v_f = (7.1 \text{ m/s}) + (-1.0 \text{ m/s}^2)(3.84 \text{ s})$$

$$v_f = 3.26 \text{ m/s}$$

(c) The range of the arrow can be calculated as follows:

$$\Delta x = v_x t$$

$$\Delta x = (3.26 \text{ m/s})(3.84 \text{ s})$$

$$\Delta x = 12.5 \text{ m}$$

(d) Without the wind, the arrow would travel in the x direction with a constant velocity of 7.1 m/s. So, the range can be calculated as follows:

$$\Delta x = v_x t$$

$$\Delta x = (7.1 \text{ m/s})(3.84 \text{ s})$$

$$\Delta x = 27.3 \text{ m}$$

Therefore, in the absence of a head wind, the arrow would travel 14.8 m farther.

120. (a) The free-body diagram of this situation looks like this.

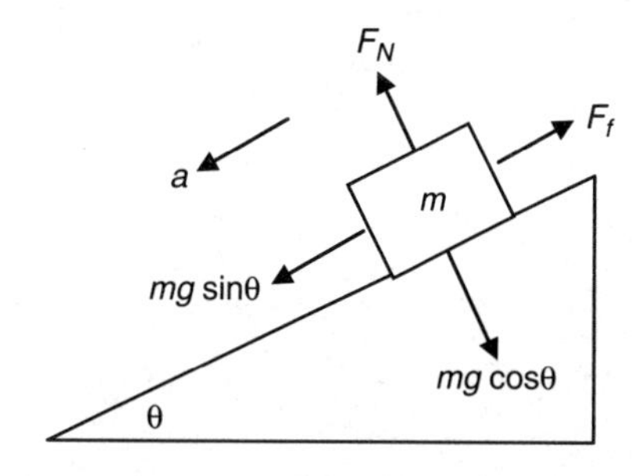

(b) We can calculate the force of friction by applying Newton's second law:

$$F_{nety} = 0$$

$$F_N - mg\cos\theta = 0$$

$$F_N = mg\cos\theta$$

$$F_f = \mu F_N$$

$$F_f = \mu mg\cos\theta$$

$$F_f = (0.1)(10 \text{ kg})(9.8 \text{ m/s}^2)\cos 30°$$

$$F_f = (0.1)(10 \text{ kg})(9.8 \text{ m/s}^2)(0.87)$$

$$F_f = 8.49 \text{ N}$$

(c) The box's acceleration can be calculated by applying Newton's second law:

$$F_{nety} = 0$$

$$F_N - mg\cos\theta = 0$$

$$F_N = mg\cos\theta$$

$$F_{netx} = ma$$

$$mg\sin\theta - F_f = ma$$

$$mg\sin\theta - \mu F_N = ma$$

$$mg\sin\theta - \mu(mg\cos\theta) = ma$$

$$g\sin\theta - \mu(g\cos\theta) = a$$

$$a = g(\sin\theta - \mu\cos\theta)$$

$$a = (9.8\ \text{m/s}^2)(\sin 30° - (0.1)\cos 30°)$$

$$a = (9.8\ \text{m/s}^2)\{(0.5) - (0.1)(0.87)\}$$

$$a = 4.05\ \text{m/s}^2$$

(d) The box started from rest and accelerated down the plane for 2 s. You can calculate the final velocity:

$$v_f = v_0 + at$$

$$v_f = (0) + (4.05\ \text{m/s}^2)(2\ \text{s})$$

$$v_f = 8.1\ \text{m/s}$$

(e) You can calculate the distance that the box moved:

$$v_f^2 = v_0^2 + 2a\Delta x$$

$$(8.1\ \text{m/s})^2 = (0\ \text{m/s})^2 + 2(4.05\ \text{m/s}^2)\Delta x$$

$$\Delta x = 8.1\ \text{m}$$

or

$$\Delta x = v_0 t + \frac{1}{2}at^2$$

$$\Delta x = 0 + \frac{1}{2}(4.05\ \text{m/s}^2)(2\ \text{s})^2$$

$$\Delta x = 8.1\ \text{m}$$

Chapter 5: Momentum

121. **(E)** The mass of the man is 70 kg and his velocity is 2 m/s. You can calculate his momentum as follows:

$$p = mv$$
$$p = (70\text{ kg})(2\text{ m/s})$$
$$p = 140\text{ kg}\cdot\text{m/s}$$

122. **(A)** The force (F) applied to the hockey puck is 10 N. The time interval (Δt) is 0.1 s. So the impulse, which is equal to the change in momentum (Δp), can be calculated as follows:

$$\Delta p = F\Delta t$$
$$\Delta p = (10\text{ N})(0.1\text{ s})$$
$$\Delta p = 1.0\text{ kg}\cdot\text{m/s}$$

123. **(A)** The mass of the car is 1,000 kg. It's moving at a constant velocity ($v = 11.0$ m/s). It comes to a complete stop. So, the change in velocity ($\Delta v = -11$ m/s) occurs over a time interval ($\Delta t = 2.0$ s). You can use the impulse–momentum theory to calculate the force acting upon the car:

$$\Delta p = F\Delta t$$
$$\Delta p = m\Delta v$$
$$F\Delta t = m\Delta v$$
$$F = \frac{m\Delta v}{\Delta t}$$
$$F = \frac{(1{,}000\text{ kg})(-11\text{ m/s})}{(2.0\text{ s})}$$
$$F = -5{,}500\text{ N}$$

124. **(B)** The ball has a mass of 1.0 kg. It moves at 10 m/s. When it hits the wall, the collision is perfectly elastic. So, the ball's direction of motion is changed and its velocity is now 10 m/s. You can calculate the ball's momentum as follows:

$$p = mv$$
$$p = (1.0\text{ kg})(-10\text{ m/s})$$
$$p = -10\text{ kg}\cdot\text{m/s}$$

125. **(C)** The mass of the box is 10 kg and the box's change in velocity is −10 m/s. The force applied to the box is constant at −10 N. Use the impulse–momentum theory to calculate the time it takes the box to stop:

$$\Delta p = F\Delta t$$
$$\Delta p = m\Delta v$$
$$F\Delta t = m\Delta v$$
$$\Delta t = \frac{m\Delta v}{F}$$
$$\Delta t = \frac{(10 \text{ kg})(-10 \text{ m/s})}{(-10 \text{ N})}$$
$$\Delta t = 10 \text{ s}$$

126. **(D)** The balls are of equal mass ($m_A = m_B$). Initially, Ball A moves at a velocity (v_A) of 10 m/s, while Ball B moves at a velocity (v_B) of −5 m/s. After the collision, Ball B moves with a velocity (v'_B) of 3 m/s. You can calculate the final velocity of Ball A by conservation of momentum in a perfectly elastic collision:

$$p_A + p_B = p'_A + p'_B$$
$$m_A v_A + m_B v_B = m_A v'_A + m_B v'_B$$
$$m_A = m_B$$
$$v_A + v_B = v'_A + v'_B$$
$$v'_A = v_A + v_B - v'_B$$
$$v'_A = (10 \text{ m/s}) + (-5 \text{ m/s}) - (3 \text{ m/s})$$
$$v'_A = 2 \text{ m/s}$$

127. **(D)** The mass of the fullback (m_A) is 140 kg, while the mass of the defender (m_B) is 70 kg. Initially, A is moving at 10 m/s and B at −5 m/s. They collide inelastically because the defender wraps his arms around the fullback and they travel together. So, you can use conservation of momentum for inelastic collisions to find their combined velocity after contact:

$$p_A + p_B = p'_{(A+B)}$$
$$m_A v_A + m_B v_B = (m_A + m_B)v'_{(A+B)}$$
$$v'_{(A+B)} = \frac{m_A v_A + m_B v_B}{(m_A + m_B)}$$
$$v'_{(A+B)} = \frac{(140 \text{ kg})(10 \text{ m/s}) + (70 \text{ kg})(-5 \text{ m/s})}{(140 \text{ kg} + 70 \text{ kg})}$$
$$v'_{(A+B)} = 5 \text{ m/s}$$

128. **(E)** The mass of Ball A equals the mass of Ball B, which is 200 g ($m_A = m_B = 0.2$ kg). Initially, Ball A travels at a velocity (v_A) of 1.0 m/s, while Ball B is at rest ($V_B = 0$ m/s). After the collision, Ball A is at rest ($V'_A = 0$ m/s). The collision is perfectly elastic, so you can calculate Ball B's velocity after the collision by conservation of momentum:

$$p_A + p_B = p'_A + p'_B$$
$$m_A v_A + m_B v_B = m_A v'_A + m_B v'_B$$
$$m_A = m_B$$
$$v_A + v_B = v'_A + v'_B$$
$$v_B = v'_A = 0$$
$$v_A = v'_B$$
$$v'_B = 1.0 \text{ m/s}$$

129. (E) The bullet's mass was 50 g (0.05 kg). Its momentum was 25 kg m/s. You can calculate the bullet's velocity as follows:

$$p = mv$$
$$v = \frac{p}{m}$$
$$v = \frac{(25 \text{ kg} \cdot \text{m/s})}{(0.05 \text{ kg})}$$
$$v = 500 \text{ m/s or } 5 \times 10^2 \text{ m/s}$$

130. (C) Object 1 (m_1 = 10 kg, x_1 = 2.0 m) and Object 2 (m_1= 20 kg, x_2 = 6.0 m) move around a common reference point. You can calculate the center of mass (x_{cm}) of the system as follows:

$$x_{cm} = \frac{m_1 x_1 + m_2 x_2}{m_1 + m_2}$$
$$x_{cm} = \frac{[(10 \text{ kg})(2.0 \text{ m}) + (20 \text{ kg})(6.0 \text{ m})]}{(10 \text{ kg} + 20 \text{ kg})}$$
$$x_{cm} = 4.7 \text{ m}$$

131. (A) The two masses (m_A, m_B) are 150 kg and 50 kg, respectively. The velocity of the halfback (v_A) is 5 m/s. You must calculate the velocity that the receiver (v_B) must have to stop the two of them after contact (v'_{A+B}= 0). You can do this by using the conservation of momentum for inelastic collisions:

$$p_A + p_B = p'_{(A+B)}$$
$$m_A v_A + m_B v_B = (m_A + m_B) v'_{(A+B)}$$
$$v'_{(A+B)} = 0$$
$$m_A v_A + m_B v_B = 0$$
$$m_B v_B = -m_A v_A$$
$$v_B = \frac{-m_A v_A}{m_B}$$
$$v_B = \frac{-(150 \text{ kg})(5 \text{ m/s})}{(50 \text{ kg})}$$
$$v_B = -15 \text{ m/s,}$$

Yes, the receiver can stop the halfback.

132. (B) The bat exerts a force of 10.0 N over a time of $\Delta t = 0.005$ s when it strikes a 145-g baseball ($m = 0.145$ kg). You can calculate the baseball's change in velocity by the impulse–momentum theory:

$$\Delta p = F\Delta t$$
$$\Delta p = m\Delta v$$
$$F\Delta t = m\Delta v$$
$$\Delta v = \frac{F\Delta t}{m}$$
$$\Delta v = \frac{(10.0\text{ N})(0.005\text{s})}{(0.145\text{ kg})}$$
$$\Delta v = 0.345\text{ m/s}$$

133. (D) The mass of an electron is 9.1×10^{-31} kg. The speed of light is 3.0×10^{8} m/s, so 90% of the speed of light is 2.7×10^{8} m/s. The momentum can be calculated as follows:

$$p = mv$$
$$p = (9.1\times 10^{-31}\text{ kg})(2.7\times 10^{8}\text{ m/s})$$
$$p = 2.5\times 10^{-22}\text{ kg}\cdot\text{m/s}$$

134. (D) The mass of the cue ball is 260 g ($m_A = 0.26$ kg), and the mass of the numbered ball is 150 g ($m_B = 0.15$ kg). The initial velocity of the cue ball is 1.0 m/s and the final velocity is zero. The initial velocity of the numbered ball is zero. The velocity of the numbered ball after the elastic collision can be found by conservation of momentum:

$$p_A + p_B = p'_A + p'_B$$
$$m_A v_A + m_B v_B = m_A v'_A + m_B v'_B$$
$$v_B = v'_A = 0$$
$$m_A v_A = m_B v'_B$$
$$v'_B = \frac{m_A v_A}{m_B}$$
$$v'_B = \frac{(0.26\text{ kg})(1.0\text{ m/s})}{(0.15\text{ kg})}$$
$$v'_B = 1.7\text{ m/s}$$

135. (A) The mass of the aircraft is 750 kg. The plane's velocity increases from 100 m/s to 120 m/s ($\Delta v = 20$ m/s). The wind blows for 2 min ($\Delta t = 120$ s). You can find the force by the impulse–momentum theorem:

$$\Delta p = F\Delta t$$
$$\Delta p = m\Delta v$$
$$F\Delta t = m\Delta v$$
$$F = \frac{m\Delta v}{\Delta t}$$
$$F = \frac{(750\text{ kg})(20\text{ m/s})}{(120\text{ s})}$$
$$F = 125\text{ N}$$

136. (C) The first derivative of a momentum–time graph is force. This is the slope of the line. In this case the slope of the line is 2. So, the force on the box is a constant 2 N.

137. (B) The force was 2.0 N and was applied for 100 m/s ($\Delta t = 0.1$ s). The object's change in velocity (Δv) was 1.0 m/s. You can calculate the mass of the object by the impulse–momentum theorem:

$$\Delta p = F\Delta t$$

$$\Delta p = m\Delta v$$

$$F\Delta t = m\Delta v$$

$$m = \frac{F\Delta t}{\Delta v}$$

$$m = \frac{(2.0 \text{ N})(0.1 \text{ s})}{(1.0 \text{ m/s})}$$

$$m = 0.2 \text{ kg}$$

138. (D) To answer this you must find the 70-kg stuntman's velocity when he hits the air bag. He has been free falling for 5 s, so his $v \approx 5t \approx 25$ m/s. He comes to a complete stop ($\Delta v =$ 25 m/s) after a time interval of 2 s ($\Delta t = 2$ s). You can use the impulse–momentum theorem:

$$\Delta p = F\Delta t$$

$$\Delta p = m\Delta v$$

$$F\Delta t = m\Delta v$$

$$F = \frac{m\Delta v}{\Delta t}$$

$$F = \frac{(70 \text{ kg})(25 \text{ m/s})}{(2 \text{ s})}$$

$$F = 875 \text{ N}$$

139. (B) Two railroad cars ($m_A = m_B = 2 \times 10^4$ kg) are traveling in the same direction along a railroad track. The cars are moving at different velocities ($v_A = 14$ m/s, $v_B = 10$ m/s). The collision is inelastic and the velocity of the combined cars (v'_{A+B}) can be calculated by conservation of momentum:

$$p_A + p_B = p'_{(A+B)}$$

$$m_A v_A + m_B v_B = (m_A + m_B)v'_{(A+B)}$$

$$v'_{(A+B)} = \frac{m_A v_A + m_B v_B}{(m_A + m_B)}$$

$$v'_{(A+B)} = \frac{(2\times10^4 \text{ kg})(14 \text{ m/s}) + (2\times10^4 \text{ kg})(10 \text{ m/s})}{(2\times10^4 \text{ kg} + 2\times10^4 \text{ kg})}$$

$$v'_{(A+B)} = 12 \text{ m/s}$$

140. (C) The first derivative of a momentum–time graph is the force on the object. In this case, the derivative is a straight line with a positive slope. So, the force on the object is increasing at a constant rate.

141. (B) The masses of the skaters are 70 kg (m_A) and 50 kg (m_B), respectively. The total momentum before the push-off was 0. The velocity (v_B) of the woman after the push-off is 2.5 m/s. You can calculate the man's velocity by the conservation of momentum:

$$p_A + p_B = 0$$

$$m_A v_A + m_B v_B = 0$$

$$m_A v_A = -m_B v_B$$

$$v_A = \frac{-m_B v_B}{m_A}$$

$$v_A = \frac{-(50\text{ kg})(2.5\text{ m/s})}{(70\text{ kg})}$$

$$v_A = -1.8\text{ m/s}$$

142. (D) The object's momentum is 10 kg·m/s and its mass is 0.5 kg. You can find the object's velocity from the definition of momentum equation:

$$p = mv$$

$$v = \frac{p}{m}$$

$$v = \frac{(10\text{ kg·m/s})}{(0.5\text{ kg})}$$

$$v = 20\text{ m/s}$$

143. (E) A cue ball ($m_A = 250$, g = 0.25 kg) travels at 1.0 m/s ($V_x = 1.0$ m/s, $V_y = 0$ m/s) and hits a numbered ball ($m_B = 170$, g = 0.17 kg) at rest ($V_x = V_y = 0$ m/s). The balls move off at angles. The numbered ball moves off at 45°, while the cue ball moves off at −45°. You need to apply the law of conservation of momentum to the *x* components and the *y* components of momentum. You will get two equations that you can solve simultaneously to get each ball's velocity:

$$p_{Ax} + p_{Bx} = p'_{Ax} + p'_{Bx}$$

$$m_A v_{Ax} + 0 = m_A v'_{Ax} + m_B v'_{Bx}$$

$$m_A v_{Ax} = m_A v'_A \cos(-45^\circ) + m_B v'_B \cos(45^\circ)$$

$$(0.25\text{ kg})(1.0\text{ m/s}) = (0.25\text{ kg})v'_A(0.707) + (0.17\text{ kg})v'_B(0.707)$$

$$(0.25\text{ kg·m/s}) = (0.177\text{ kg})v'_A + (0.12\text{ kg})v'_B$$

$$p_{Ay} + p_{By} = p'_{Ay} + p'_{By}$$

$$0 + 0 = m_A v'_{Ay} + m_B v'_{By}$$

$$0 = m_A v'_A \sin(-45^\circ) + m_B v'_B \sin(45^\circ)$$

$$0 = (0.25\text{ kg})v'_A(-0.707) + (0.17\text{ kg})v'_B(0.707)$$

$$0 = (-0.177\text{ kg})v'_A + (0.12\text{ kg})v'_B$$

Add the two equations:

$(0.25 \text{ kg} \cdot \text{m/s}) = (0.177 \text{ kg})v'_A + (0.12 \text{ kg})v'_B$

$0 = (-0.177 \text{ kg})v'_A + (0.12 \text{ kg})v'_B$

$(0.25 \text{ kg} \cdot \text{m/s}) = (0.24 \text{ kg})v'_B$

$$v'_B = \frac{(0.25 \text{ kg} \cdot \text{m/s})}{(0.24 \text{ kg})}$$

$v'_B = 1.04 \text{ m/s}$, Now substitute this value into one of the above equations:

$0 = (-0.177 \text{ kg})v'_A + (0.12 \text{ kg})(1.04 \text{ m/s})$

$(0.177 \text{ kg})v'_A = (0.125 \text{ kg} \cdot \text{m/s})$

$$v'_A = \frac{(0.125 \text{ kg} \cdot \text{m/s})}{(0.177 \text{ kg})}$$

$v'_A = 0.71 \text{ m/s}$

144. (E) The mass of the airplane is 2,000 kg and the velocity is 343 m/s. You can calculate the momentum:

$p = mv$

$p = (2000 \text{ kg})(343 \text{ m/s})$

$p = 6.9 \times 10^5 \text{ kg} \cdot \text{m/s}$

145. (A) The mass of the handgun (m_A) is 1.2 kg and the mass of the bullet (m_B) is 7.5 g (0.0075 kg). The bullet travels away at a velocity (v_B) of 365 m/s. You can calculate the recoil velocity of the handgun by conservation of momentum:

$0 = p_A + p_B$

$0 = m_A v_A + m_B v_B$

$m_A v_A = -m_B v_B$

$$v_A = \frac{-m_B v_B}{m_A}$$

$$v_A = \frac{-(0.0075 \text{ kg})(365 \text{ m/s})}{(1.2 \text{ kg})}$$

$v_A = -2.4 \text{ m/s}$

146. (D) The two cars have equal masses ($m_A = m_B = 0.5$ kg). Car A moves at 0.1 m/s and Car B is at rest ($v_B = 0$ m/s). The two cars collide and interlock. This is an inelastic collision. You can find the velocity of the combined cars (v'_{A+B}) by conservation of momentum:

$$p_A + p_B = p'_{(A+B)}$$

$$m_A v_A + 0 = (m_A + m_B) v'_{(A+B)}$$

$$v'_{(A+B)} = \frac{m_A v_A}{(m_A + m_B)}$$

$$v'_{(A+B)} = \frac{(0.5 \text{ kg})(0.1 \text{ m/s})}{(0.5 \text{ kg} + 0.5 \text{ kg})}$$

$$v'_{(A+B)} = 0.05 \text{ m/s}$$

147. (C) The bullet's mass is 4 g (0.004 kg) and its velocity is 950 m/s. The time interval that the rifle exerts on the bullet is ($\Delta t = 0.1$ s). You can find the force from the impulse–momentum theorem:

$$\Delta p = F \Delta t$$

$$\Delta p = m \Delta v$$

$$F \Delta t = m \Delta v$$

$$F = \frac{m \Delta v}{\Delta t}$$

$$F = \frac{(0.004 \text{ kg})(950 \text{ m/s})}{(0.1 \text{ s})}$$

$$F = 38 \text{ N}$$

148. (C) The masses of the cannon ($m_A = 1{,}000$ kg) and cannonball ($m_B = 15$ kg) are known. The recoil velocity of the cannon (v_A) is –1.5 m/s. You can calculate the velocity of the cannonball (v_B) by conservation of momentum:

$$0 = p_A + p_B$$

$$0 = m_A v_A + m_B v_B$$

$$m_B v_B = -m_A v_A$$

$$v_B = \frac{-m_A v_A}{m_B}$$

$$v_B = \frac{-(1{,}000 \text{ kg})(-1.5 \text{ m/s})}{(15 \text{ kg})}$$

$$v_B = 100 \text{ m/s}$$

149.

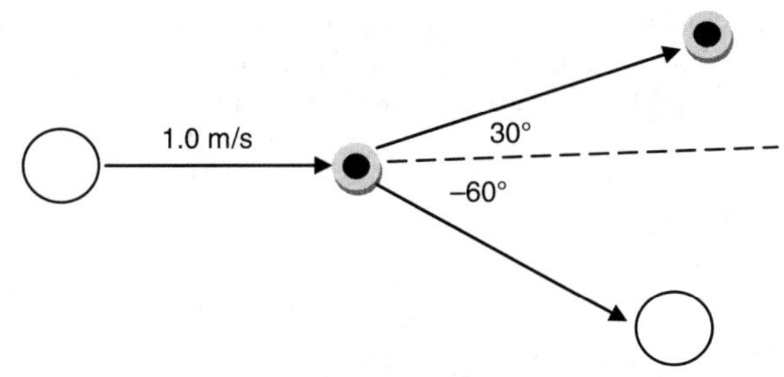

$p_{Ax} + p_{Bx} = p'_{Ax} + p'_{Bx}$

$m_A v_{Ax} + 0 = m_A v'_{Ax} + m_B v'_{Bx}$

$m_A v_{Ax} = m_A v'_A \cos(-60^\circ) + m_B v'_B \cos(30^\circ)$

$(0.25 \text{ kg})(1.0 \text{ m/s}) = (0.25 \text{ kg})v'_A(0.5) + (0.17 \text{ kg})v'_B(0.87)$

$(0.25 \text{ kg}\cdot\text{m/s}) = (0.125 \text{ kg})v'_A + (0.15 \text{ kg})v'_B$

$p_{Ax} + p_{By} = p'_{Ay} + p'_{By}$

$0 + 0 = m_A v'_{Ay} + m_B v'_{By}$

$0 = m_A v'_A \sin(-60^\circ) + m_B v'_B \sin(30^\circ)$

$0 = (0.25 \text{ kg})v'_A(-0.87) + (0.17 \text{ kg})v'_B(0.0.5)$

$0 = (-0.22 \text{ kg})v'_A + (0.09 \text{ kg})v'_B$

Add the two equations:

$(0.25 \text{ kg}\cdot\text{m/s}) = (0.125 \text{ kg})v'_A + (0.15 \text{ kg})v'_B$

$0 = (-0.22 \text{ kg})v'_A + (0.09 \text{ kg})v'_B$

$(0.44 \text{ kg}\cdot\text{m/s}) = (0.22 \text{ kg})v'_A + (0.264 \text{ kg})v'_B$

$0 = (-0.22 \text{ kg})v'_A + (0.09 \text{ kg})v'_B$

$(0.44 \text{ kg}\cdot\text{m/s}) = (0.354 \text{ kg})v'_B$

$$v'_B = \frac{(0.44 \text{ kg}\cdot\text{m/s})}{(0.354 \text{ kg})}$$

$v'_B = 1.24$ m/s, Now substitute this value into one of the above equations:

$0 = (-0.22 \text{ kg})v'_A + (0.09 \text{ kg})(1.24 \text{ m/s})$

$(0.22 \text{ kg})v'_A = (0.112 \text{ kg}\cdot\text{m/s})$

$$v'_A = \frac{(0.112 \text{ kg}\cdot\text{m/s})}{(0.22 \text{ kg})}$$

$v'_A = 0.51$ m/s

150. (a) Let's designate Car B as the 800-kg car (v_B= 25.5 m/s) and Car A as the 1,000-kg car (v_A = 34.7 m/s). The two cars collide in an inelastic collision. You can find the combined velocity (v'_{A+B}) by the conservation of momentum:

$$p_A + p_B = p'_{(A+B)}$$

$$m_A v_A + m_B v_B = (m_A + m_B)v'_{(A+B)}$$

$$v'_{(A+B)} = \frac{m_A v_A + m_B v_B}{(m_A + m_B)}$$

$$v'_{(A+B)} = \frac{(1{,}000 \text{ kg})(34.7 \text{ m/s}) + (800 \text{ kg})(25.5 \text{ m/s})}{(1{,}000 \text{ kg} + 800 \text{ kg})}$$

$$v'_{(A+B)} = 30.6 \text{ m/s}$$

(b) $F_f = \mu_k F_N$

$$F_N = (m_A + m_B)g$$

$$F_f = \mu_k (m_A + m_B)g$$

$$F_f = (0.7)(1{,}000 \text{ kg} + 800 \text{ kg})(9.8 \text{ m/s}^2)$$

$$F_f = 1.23 \times 10^4 \text{ N}$$

(c) The force of friction is negative because it opposes motion. You can calculate the time (Δt) it takes for the interlocked cars to stop ($\Delta v = -30.6$ m/s) by using the impulse–momentum theorem:

$$\Delta p = F\Delta t$$

$$\Delta p = m\Delta v$$

$$F\Delta t = m\Delta v$$

$$F_f \Delta t = (m_A + m_B)\Delta v$$

$$\Delta t = \frac{(m_A + m_B)\Delta v}{F_f}$$

$$\Delta t = \frac{(1{,}000 \text{ kg} + 800 \text{ kg})(-30.6 \text{ m/s})}{(-1.23 \times 10^4 \text{ N})}$$

$$\Delta t = 4.46 \text{ s}$$

Chapter 6: Energy Conservation

151. (D) The mass of the man is 70 kg and his velocity is 2 m/s. You can calculate his kinetic energy as follows:

$$KE = \frac{1}{2}mv^2$$

$$KE = \frac{1}{2}(70 \text{ kg})(2 \text{ m/s})^2$$

$$KE = 140 \text{ J}$$

152. (D) The force (F) applied to the box is 10 N in the direction of motion. The force is applied over a 10-m distance. So the work can be calculated as follows:

$$W = Fd$$

$$W = (10 \text{ N})(10 \text{ m})$$

$$W = 100 \text{ J}$$

153. (C) The mass of the car is 1,000 kg. Its initial velocity is 0 m/s. The net force is 500 N and it acts constantly over a distance (d) of 100 m. You can use the work–energy theorem to calculate the car's final velocity:

$$W = Fd$$

$$\Delta KE = \frac{1}{2}m(\Delta v)^2$$

$$W = \Delta KE$$

$$Fd = \frac{1}{2}m(\Delta v)^2$$

$$2Fd = m(\Delta v)^2$$

$$\Delta v = v_f - 0 = v_f$$

$$2Fd = mv_f^2$$

$$v_f^2 = \frac{2Fd}{m}$$

$$v_f = \sqrt{\frac{2Fd}{m}}$$

$$v_f = \sqrt{\frac{2(500 \text{ N})(100 \text{ m})}{(100 \text{ kg})}}$$

$$v_f = 10 \text{ m/s}$$

154. **(C)** The piano has a mass of 500 kg. Its change in height (h) is 10 m. You can calculate the piano's change in potential energy (ΔPE) as follows:

$$\Delta PE = mgh$$

$$\Delta PE = (500\text{ kg})(10\text{ m/s}^2)(10\text{ m})$$

$$\Delta PE = 5\times10^4\text{ J}$$

155. **(D)** The force on the box is 10 N at an angle ($\theta = 60°$) over a distance (d) of 50 m. Only the component of force in the horizontal direction contributes to the work done on the box. You must find the x component of force and use it to calculate the work:

$$W = F_x d$$

$$F_x = F\cos\theta$$

$$W = F\cos\theta d$$

$$W = (10\text{ N})(\cos 60°)(50\text{ m})$$

$$W = 250\text{ J}$$

156. **(C)** The mass of the ball is 5 kg, but this information is irrelevant. The height of the hill (h) is 10 m. The ball starts from rest and reaches an unknown final velocity (v). So the ball's change in velocity is equal to its final velocity ($\Delta v = v_f$). The ball's change in potential energy is equal to its change in kinetic energy. So, you can calculate the ball's final velocity from conservation of energy:

$$KE_i + PE_i = KE_f + PE_f$$

$$0 + PE_i = KE_f + 0$$

$$KE_f = PE_i$$

$$\frac{1}{2}mv_f^2 = mgh$$

$$\frac{1}{2}v_f^2 = gh$$

$$v_f^2 = 2gh$$

$$v_f = \sqrt{2gh}$$

$$v_f = \sqrt{2(10\text{ m/s}^2)(10\text{ m})}$$

$$v_f = 14\text{ m/s}$$

157. (C) The mass of the arrow (m) is 20 g or 0.02 kg. The archer exerts a force ($F = 450$ N) to pull the bowstring back ($d = 1.5$ m). You can calculate the final velocity of the arrow by using the work–energy theorem:

$$W = Fd$$

$$\Delta KE = \frac{1}{2}m(\Delta v)^2$$

$$W = \Delta KE$$

$$Fd = \frac{1}{2}m(\Delta v)^2$$

$$2Fd = m(\Delta v)^2$$

$$\Delta v = v_f - 0 = v_f$$

$$2Fd = mv_f^2$$

$$v_f^2 = \frac{2Fd}{m}$$

$$v_f = \sqrt{\frac{2Fd}{m}}$$

$$v_f = \sqrt{\frac{2(450 \text{ N})(1.5 \text{ m})}{(0.02 \text{ kg})}}$$

$$v_f = 260 \text{ m/s}$$

158. (E) The mass of the car is 1,000 kg. The change in velocity ($\Delta v = -30$ m/s) occurs over a distance (d) of 10 m. You can calculate the force by the work–energy theorem:

$$W = Fd$$

$$\Delta KE = \frac{1}{2}m(\Delta v)^2$$

$$W = \Delta KE$$

$$Fd = \frac{1}{2}m(\Delta v)^2$$

$$F = \frac{m(\Delta v)^2}{2d}$$

$$F = \frac{(1{,}000 \text{ kg})(-30 \text{ m/s})^2}{2(10 \text{ m})}$$

$$F = 4.5 \times 10^4 \text{ N}$$

159. (E) The bullet's mass was 7.5 g (0.0075 kg). The block's mass was 2.50 kg. It swung to a height of 0.1 m. When the bullet strikes the wood, it is an inelastic collision. The kinetic energy of the bullet and the block get converted to potential energy. You can calculate the bullet's velocity as follows:

First, calculate the initial velocity of the block and bullet by conservation of energy:

$$KE_i + KE_f = PE_i + PE_f$$

$$KE_i + 0 = 0 + PE_f$$

$$KE_i = PE_f$$

$$\frac{1}{2}(m_A + m_B)v_i^2 = (m_A + m_B)gh$$

$$\frac{1}{2}v_i^2 = gh$$

$$v_i^2 = 2gh$$

$$v_i = \sqrt{2gh}$$

$$v_i = \sqrt{2(10\text{ m/s}^2)(0.1\text{ m})}$$

$$v_i = 1.44\text{ m/s} = v'_{(A+B)}$$

Now, use conservation of momentum for an inelastic collision to calculate the bullet's initial velocity:

$$p_A + p_B = p'_{(A+B)}$$

$$m_A v_A + m_B v_B = (m_A + m_B)v'_{(A+B)}$$

$$v_B = 0$$

$$m_A v_A + 0 = (m_A + m_B)v'_{(A+B)}$$

$$m_A v_A = (m_A + m_B)v'_{(A+B)}$$

$$v_A = \frac{(m_A + m_B)v'_{(A+B)}}{m_A}$$

$$v_A = \frac{(0.0075\text{ kg} + 2.50\text{ kg})(1.44\text{ m/s})}{(0.0075\text{ kg})}$$

$$v_A = 481\text{ m/s} \simeq 500\text{ m/s}$$

160. (B) Potential energy is the integral of force on a force versus displacement graph. By using trapezoidal integration for each 0.5-m interval, you can calculate the total potential energy:

$$A = \frac{1}{2}(b_1 + b_2)h$$

$$0\text{–}0.5 \text{ m: } A = \frac{1}{2}(0 \text{ N} + 5 \text{ N})(0.5 \text{ m}) = 1.25 \text{ J}$$

$$0.5\text{–}1 \text{ m: } A = \frac{1}{2}(5 \text{ N} + 15 \text{ N})(0.5 \text{ m}) = 5.0 \text{ J}$$

$$1\text{–}1.5 \text{ m: } A = \frac{1}{2}(15 \text{ N} + 20 \text{ N})(0.5 \text{ m}) = 8.75 \text{ J}$$

$$1.5\text{–}2 \text{ m: } A = \frac{1}{2}(20 \text{ N} + 20 \text{ N})(0.5 \text{ m}) = 10 \text{ J}$$

Total = 25 J

161. (C) A spring's constant (k) is 300 N/m. The spring is stretched (x) by 0.5 m. You can calculate the force on the spring by Hooke's law:

$$F = kx$$

$$F = (300 \text{ N/m})(0.5 \text{ m})$$

$$F = 150 \text{ N}$$

162. (A) A spring's constant (k) is 400 N/m. The spring is stretched (x) by 0.5 m. You can calculate the potential energy on the spring:

$$PE = \frac{1}{2}kx^2$$

$$PE = \frac{1}{2}(400 \text{ N/m})(0.5 \text{ m})^2$$

$$PE = 50 \text{ J}$$

163. (D) The spring has a constant of 100 N/m and is compressed ($x = -0.2$ m). The mass of the block is 1 kg. You can calculate the block's velocity as it passes equilibrium by conservation of energy:

$$KE_i + PE_i = KE_f + PE_f$$

$$0 + PE_i = KE_f + 0$$

$$KE_f = PE_i$$

$$\frac{1}{2}mv_f^2 = \frac{1}{2}kx^2$$

$$mv_f^2 = kx^2$$

$$v_f^2 = \frac{kx^2}{m}$$

$$v_f = \sqrt{\frac{kx^2}{m}}$$

$$v_f = \sqrt{\frac{(100\text{ N/m})(-0.2\text{ m})^2}{(1\text{ kg})}}$$

$$v_f = 2\text{ m/s}$$

164. (C) Both cannons fire balls of the same mass (m) and use the same amount of powder to supply identical forces (F). The length of Cannon 2 (d_2) is two times longer than that of Cannon 1 (d_1). You can solve this problem by applying the work–energy theorem to both cannons:

$$Fd_1 = \frac{1}{2}mv_1^2$$

$$Fd_2 = \frac{1}{2}mv_2^2$$

$$d_2 = 2d_1$$

$$F(2d_1) = \frac{1}{2}mv_2^2$$

$$Fd_1 = \frac{1}{4}mv_2^2$$

$$\frac{1}{4}mv_2^2 = \frac{1}{2}mv_1^2$$

$$\frac{1}{4}v_2^2 = \frac{1}{2}v_1^2$$

$$v_2^2 = 2v_1^2$$

$$v_2 = \sqrt{2v_1^2}$$

$$v_2 = \sqrt{2}\sqrt{v_1^2}$$

$$v_2 = \sqrt{2}v_1 = 1.4v_1$$

165. (A) The mass of the aircraft is 750 kg. The plane's velocity increases from 100 m/s to 120 m/s ($\Delta v = 20$ m/s). The wind blows for 1,200 m (Δd). You can find the force by the work–energy theorem:

$$F\Delta d = \frac{1}{2}m\Delta v^2$$

$$F = \frac{m\Delta v^2}{2\Delta d}$$

$$F = \frac{(750\text{ kg})(20\text{ m/s})^2}{2(1{,}200\text{ m})}$$

$$F = 125\text{ N}$$

166. (D) The first derivative of a potential energy–distance graph is force. This graph has a quadratic function, so the first derivative is a straight line. From the shape of this graph, the slope of the line is positive. Therefore, the force is changing linearly with a positive slope.

167. (D) The force was 2.0 N and was applied for 5.0 m. The object's change in velocity (Δv) was 1.0 m/s. You can calculate the mass of the object by the work–energy theorem:

$$W = Fd$$

$$\Delta KE = \frac{1}{2}m(\Delta v)^2$$

$$W = \Delta KE$$

$$Fd = \frac{1}{2}m(\Delta v)^2$$

$$2Fd = m(\Delta v)^2$$

$$m = \frac{2Fd}{(\Delta v)^2}$$

$$m = \frac{2(2.0\text{ N})(5.0\text{ m})}{(1.0\text{ m/s})^2}$$

$$m = 20\text{ kg}$$

168. (C) According to conservation of energy, the potential energy of the stuntman gets converted to kinetic energy as he falls. The maximum kinetic energy occurs when he hits the airbag. The airbag must do work to stop the stuntman's kinetic energy. So, you can solve this by applying conservation of energy and the work–energy theorem to the situation:

$$KE_i + PE_i = KE_f + PE_f$$

$$0 + PE_i = KE_f + 0$$

$$KE_f = PE_i$$

$$\frac{1}{2}mv_f^2 = mgh$$

$$W = \Delta KE$$

$$Fd = \frac{1}{2}mv_f^2$$

$$Fd = mgh$$

$$F = \frac{mgh}{d}$$

$$F = \frac{(70\text{ kg})(10\text{ m/s}^2)(125\text{ m})}{(5\text{ m})}$$

$$F = 1.75 \times 10^3\text{ N}$$

169. **(E)** The force exerted by the man is 100 N in the direction of motion. The mass of the box is 100 kg, which is irrelevant to the solution. The distance is 60 m and the time interval is 2 min ($\Delta t = 120s$). You can calculate the power as follows:

$$P = \frac{W}{t}$$

$$P = \frac{Fd}{t}$$

$$P = \frac{(100 \text{ N})(60 \text{ m})}{(120 \text{ s})}$$

$$P = 50 \text{ W}$$

170. **(E)** Points A and C are at the same height and would have the same potential energy.

171. **(E)** The car will move the fastest at the point with the most kinetic energy, which is the lowest point, E.

172. **(E)** The total energy is the sum of the kinetic energy and potential energy. It is the same at all points.

173. **(A)** The man exerts a component (F_x) of the 20 N force (F) in the horizontal by applying the force at an angle ($\theta = 60°$). The lawn mower moves a horizontal distance (d) of 100 m in a time interval (Δt) of 5 min (300 s). Only the horizontal component of force does work, so you must resolve the force into components and use that to calculate the power:

$$P = \frac{W}{\Delta t}$$

$$W = F_x d$$

$$F_x = F\cos\theta$$

$$W = F\cos\theta d$$

$$P = \frac{F\cos\theta d}{\Delta t}$$

$$P = \frac{(20 \text{ N})(\cos 60°)(100 \text{ m})}{(300 \text{ s})}$$

$$P = 3 \text{ W}$$

174. **(D)** The mass of the airplane is 2,000 kg and the velocity is 343 m/s. You can calculate the kinetic energy:

$$KE = \frac{1}{2}mv^2$$

$$KE = \frac{1}{2}(2{,}000 \text{ kg})(343 \text{ m/s})^2$$

$$KE = 1.18 \times 10^8 \text{ J}$$

175. (D) The handgun does work on the bullet and gives it kinetic energy. All the kinetic energy ($\Delta KE = 500$ J) is lost as heat. The length of the gun barrel is 125 mm ($d = 0.125$ m). You can calculate the force that the handgun exerts on the bullet using the work–energy theorem:

$$W = \Delta KE$$

$$Fd = \Delta KE$$

$$F = \frac{\Delta KE}{d}$$

$$F = \frac{(500 \text{ J})}{(0.125 \text{ m})}$$

$$F = 4{,}000 \text{ N}$$

176. (D) The two cars have equal masses ($m_A = m_B = 0.5$ kg). Car A moves at 0.2 m/s and Car B is at rest ($v_B = 0$ m/s). The two cars collide in an elastic collision. Both cars move away at 0.1 m/s. You must calculate the kinetic energies of the cars before and after the collision. Since you know that energy must be conserved, the difference between the kinetic energies before and after represents energy lost as heat. (No cars have potential energy because they are at the same height on the track.)

$$KE_A + KE_B = KE'_A + KE'_B + heat$$

$$KE_A + 0 = KE'_A + KE'_B + heat$$

$$KE_A = KE'_A + KE'_B + heat$$

$$heat = KE_A - KE'_A - KE'_B$$

$$heat = \frac{1}{2}m_A v_A^2 - \frac{1}{2}m_A v_A'^2 - \frac{1}{2}m_B v_B'^2$$

$$m_A = m_B$$

$$heat = \frac{1}{2}m_A v_A^2 - \frac{1}{2}m_A v_A'^2 - \frac{1}{2}m_A v_B'^2$$

$$heat = \frac{1}{2}m_A\left(v_A^2 - v_A'^2 - v_B'^2\right)$$

$$heat = \frac{1}{2}(0.5 \text{ kg})\{(0.2 \text{ m/s})^2 - (0.1 \text{ m/s})^2 - (0.1 \text{ m/s})^2\}$$

$$heat = 0.005 \text{ J}$$

$$KE_A = \frac{1}{2}m_A v_A^2 = \frac{1}{2}(0.5 \text{ kg})(0.2 \text{ m/s})^2 = 0.01 \text{ J}$$

$$\frac{heat}{KE_A} = \frac{(0.005 \text{ J})}{(0.01 \text{ J})} = 0.5 = 50\%$$

177. (C) The bullet's mass is 4 g (0.004 kg) and its velocity is 950 m/s. The length of the rifle barrel is 1.01 m. You can find the force from the work–energy theorem:

$$W = \Delta KE$$

$$Fd = \frac{1}{2}m\Delta v^2$$

$$F = \frac{m\Delta v^2}{2d}$$

$$F = \frac{(0.004\text{ kg})(950\text{ m/s})^2}{2(1.01\text{ m})}$$

$$F = 1{,}800\text{ N}$$

178. (C) The mass of the cannon is irrelevant. The mass of the cannonball is known ($m = 11$ kg). The powder charge exerts a force of 2.24×10^4 N over the length of the cannon's barrel ($d = 2.44$ m). You can calculate the velocity of the cannonball (v) by using the work–energy theorem:

$$W = \Delta KE$$

$$Fd = \frac{1}{2}mv^2$$

$$2Fd = mv^2$$

$$v^2 = \frac{2Fd}{m}$$

$$v = \sqrt{\frac{2Fd}{m}}$$

$$v = \sqrt{\frac{2(2.25\times 10^4\text{ N})(2.44\text{ m})}{(11\text{ kg})}}$$

$$v = 100\text{ m/s}$$

179. (a) You must find the height of the ramp and use conservation of energy to find the velocity of the ball:

$$KE_i + PE_i = KE_f + PE_f$$

$$0 + PE_i = KE_f + 0$$

$$KE_f = PE_i$$

$$\frac{1}{2}mv_f^2 = mgh$$

$$\sin\theta = \frac{h}{d}$$

$$h = d\sin\theta$$

$$\frac{1}{2}v_f^2 = gd\sin\theta$$

$$v_f^2 = 2gd\sin\theta$$

$$v_f = \sqrt{2gd\sin\theta}$$

$$v_f = \sqrt{2(10\text{ m/s}^2)(10\text{ m})(\sin 45°)}$$

$$v_f = 12\text{ m/s}$$

(b) Now that you know the velocity, you can calculate the ball's kinetic energy:

$$KE = \frac{1}{2}mv^2$$

$$KE = \frac{1}{2}(5\text{ kg})(12\text{ m/s})^2$$

$$KE = 360\text{ J}$$

(c) The force of friction works on the ball constantly to bring it to a stop. You must first calculate the force of friction and then use the work–energy theorem to find the distance the ball travels:

$$W = F_f d$$

$$\Delta KE = \frac{1}{2}m(\Delta v)^2$$

$$W = \Delta KE$$

$$F_f d = \frac{1}{2}m(\Delta v)^2$$

$$F_f = \mu F_N$$

$$F_N = mg$$

$$F_f = \mu mg$$

$$2\mu mgd = m(\Delta v)^2$$

$$2\mu gd = (\Delta v)^2$$

$$d = \frac{(\Delta v)^2}{2\mu g}$$

$$d = \frac{(12\text{ m/s})^2}{2(0.6)(10\text{ m/s}^2)}$$

$$d = 12\text{ m}$$

(d) You know the change in velocity of the ball and the force of friction, so you can calculate the time it takes to stop by the impulse–momentum theorem:

$$\Delta p = F\Delta t$$

$$\Delta p = m\Delta v$$

$$F\Delta t = m\Delta v$$

$$F_f\Delta t = m\Delta v$$

$$F_f = \mu F_N = \mu mg$$

$$\mu mg\Delta t = m\Delta v$$

$$\mu g\Delta t = \Delta v$$

$$\Delta t = \frac{\Delta v}{\mu g}$$

$$\Delta t = \frac{(12 \text{ m/s})}{(0.6)(10 \text{ m/s}^2)}$$

$$\Delta t = 2.0 \text{ s}$$

180. (a) You can determine the change in velocity of the ship by performing the following calculations:

$$v_i = \left(\frac{10 \text{ km}}{\text{h}}\right)\left(\frac{1{,}000 \text{ m}}{\text{km}}\right)\left(\frac{\text{h}}{3{,}600 \text{ s}}\right) = 2.78 \text{ m/s}$$

$$v_f = \left(\frac{13 \text{ km}}{\text{h}}\right)\left(\frac{1{,}000 \text{ m}}{\text{km}}\right)\left(\frac{\text{h}}{3{,}600 \text{ s}}\right) = 3.61 \text{ m/s}$$

$$\Delta v = v_f - v_i$$

$$\Delta v = (3.61 \text{ m/s}) - (2.78 \text{ m/s})$$

$$\Delta v = 0.83 \text{ m/s}$$

(b) By knowing the change in velocity and the mass of the ship, you can calculate the ship's change in kinetic energy:

$$\Delta KE = \frac{1}{2}m(\Delta v)^2$$

$$\Delta KE = \frac{1}{2}(1.43\times10^6 \text{ kg})(0.83 \text{ m/s})^2$$

$$\Delta KE = 4.93\times10^5 \text{ J}$$

(c) First, you have to calculate the component of the force of the wind acting in the ship's direction of motion. Then, use that to calculate the force of the wind:

$$W = F_y d$$

$$\Delta KE = \frac{1}{2}m(\Delta v)^2$$

$$W = \Delta KE$$

$$F_y d = \frac{1}{2}m(\Delta v)^2$$

$$F_y = F \sin\theta$$

$$F \sin\theta d = \frac{1}{2}m(\Delta v)^2$$

$$F = \frac{m(\Delta v)^2}{2d \sin\theta}$$

$$F = \frac{(1.43\times10^6 \text{ kg})(0.83 \text{ m/s})^2}{2(1{,}000 \text{ m})(\sin 60^\circ)}$$

$$F = 570 \text{ N}$$

Chapter 7: Gravitation and Circular Motion

181. (C) You are given the tangential velocity of 30 m/s and the radius of the circle of 10 m. You are asked to calculate the centripetal acceleration, which is toward the center.

$$a_c = \frac{v^2}{r}$$

$$a_c = \frac{(30 \text{ m/s})^2}{(10 \text{ m})}$$

$$a_c = 90 \text{ m/s}^2$$

182. (D) The force of friction is what holds the car in the circular motion and is equal to the centripetal force. You calculated the centripetal acceleration in **Question 181**. By knowing the mass of the car (1,000 kg), you can now calculate the centripetal force, which acts toward the center.

$$F_c = ma_c$$

$$F_c = (1{,}000 \text{ kg})(90 \text{ m/s}^2)$$

$$F_c = 90{,}000 \text{ N}$$

183. (D) You know the masses of the satellite (100 kg) and Earth (6×10^{24} kg) and the radius of the satellite's orbit (100 km + radius of the Earth = 100,000 m + 6.4×10^6 m = 6.5×10^6 m). The value of G (6.7×10^{-11} N·m²/kg²) is implied, but it is not stated. Use Newton's law of gravitation to find the force of gravity:

$$F_g = \frac{Gm_1m_2}{r^2}$$

$$F_g = \frac{(6.7\times10^{-11}\ \text{N}\cdot\text{m}^2/\text{kg}^2)(100\ \text{kg})(6\times10^{24}\ \text{kg})}{(6.5\times10^6\ \text{m})^2}$$

$$F_g = 6.2\times10^9\ \text{N}$$

184. (E) You must compare the gravitational force on Satellite A to Satellite B using the equations of Newton's law of universal gravitation. Note that the radius of Satellite B is twice that of Satellite A.

$$F_{gA} = \frac{Gm_Am_p}{r_A^2}$$

$$F_{gB} = \frac{Gm_Bm_p}{r_B^2}$$

$$r_B = 2r_A \text{ and } m_B = m_A$$

$$F_{gB} = \frac{Gm_Am_p}{(2r_A)^2}$$

$$F_{gB} = \frac{Gm_Am_p}{4r_A^2}$$

$$\frac{F_{gA} = \dfrac{Gm_Am_p}{r_A^2}}{F_{gB} = \dfrac{Gm_Am_p}{4r_A^2}}$$

$$\frac{F_{gA}}{F_{gB}} = 4$$

$$F_{gA} = 4F_{gB}$$

185. (A) You know the mass of the astronaut (70 kg), the mass of the spacecraft (5×10^4 kg), and the distance between them (10 m). Use Newton's law of gravitation to calculate the force of attraction between them:

$$F_g = \frac{Gm_1m_2}{r^2}$$

$$F_g = \frac{(6.7\times10^{-11}\ \text{N}\cdot\text{m}^2/\text{kg}^2)(70\ \text{kg})(5\times10^4\ \text{kg})}{(10\ \text{m})^2}$$

$$F_g = 2.4\times10^{-6}\ \text{N}$$

186. (A) The centripetal acceleration on the car is 1×10^5 m/s^2 and the radius of the turn is 10 m. You can use the centripetal acceleration equation to find the car's velocity:

$$a_c = \frac{v^2}{r}$$

$$v^2 = a_c r$$

$$v = \sqrt{a_c r}$$

$$v = \sqrt{\frac{(1\times10^5 \text{ m/s}^2)}{(1{,}000 \text{ kg})}}$$

$$v = 1\times10^1 \text{ m/s}$$

187. (A) You know the diameter of the ice rink (20 m), so you can calculate its circumference. From the circumference and the time it takes her to go around the rink once (62.8 s), you can calculate the velocity. Using the velocity, the radius, and acceleration due to gravity, you can calculate the coefficient of friction of the ice:

$$v = \frac{2\pi r}{t}$$

$$v = \frac{2(3.14)(10 \text{ m})}{(62.8 \text{ s})}$$

$$v = 1 \text{ m/s}$$

$$\mu = \frac{v^2}{gr}$$

$$\mu = \frac{(1 \text{ m/s})^2}{(10 \text{ m/s}^2)(10 \text{ m})}$$

$$\mu = 0.01$$

188. (C) You know the mass of the payload (1,000 kg), and the Earth (6×10^{24} kg), the Earth's radius (6.4×10^6 m), and the height of the orbit (150 km). You can calculate the gravitational potential energy of the payload when it reaches orbit:

$$PE_g = \frac{Gm_1m_2}{r}$$

$$PE_g = \frac{(6.7\times10^{-11} \text{ N}\cdot\text{m}^2/\text{kg}^2)(1000 \text{ kg})(6\times10^{24} \text{ kg})}{(6.4\times10^6 \text{ m}+1.5\times10^5 \text{ m})}$$

$$PE_g = 6.1\times10^{10} \text{ J}$$

189. (B) The warrior spins a slingshot in a horizontal circle above his head at a constant velocity. When it is released, the stone will fly off at that velocity. You know the stone's mass (50 g = 0.05 kg) and the sling's radius (1.5 m). The tension in the string is equal to the centripetal force (3.3 N), so you can calculate the velocity:

$$F_c = \frac{mv^2}{r}$$

$$v^2 = \frac{F_c r}{m}$$

$$v = \sqrt{\frac{F_c r}{m}}$$

$$v = \sqrt{\frac{(3.3\ \text{N})(1.5\ \text{m})}{(0.05\ \text{kg})}}$$

$$v = 10\ \text{m/s}$$

190. (C) An acceleration of 1 G is approximately 10 m/s^2. So a "3-G acceleration" would be approximately 30 m/s^2. On the graph this corresponds to a radius of approximately 3 m.

191. (C) From the graph, you can see that a radius of 5 m results in an acceleration of 20 m/s^2, or approximately 2 Gs. Since you know the astronaut's mass (70 kg), you can calculate the magnitude of the centripetal force:

$$F_c = ma_c$$

$$F_c = (70\ \text{kg})(20\ \text{m/s}^2)$$

$$F_c = 140\ \text{N}$$

192. (D) Kepler's third law states that the square of the period of a planet's orbit (T) is proportional to the distance from the Sun (a) cubed. When the period is expressed in Earth years and the orbital radius in AU, then the law is $T^2 = a^3$. You know that Jupiter's orbital distance is 5 AU, so you can use Kepler's third law to calculate its orbital period:

$$T^2 = a^3$$

$$T = \sqrt{a^3}$$

$$T = \sqrt{(5\ \text{AU})^3}$$

$$T = 11\ \text{years}$$

193. (C) You know the satellite's altitude (200 km), the mass of the Earth (6×10^{24} kg), and the Earth's radius (6.4×10^6 m). You must convert the satellite's altitude to *m* and add the Earth's radius to it to get the satellite's orbital radius. Then, you can calculate the satellite's velocity:

$$v = \sqrt{\frac{GM_E}{r}}$$

$$v = \sqrt{\frac{(6.7 \times 10^{-11}\ \text{N} \cdot \text{m}^2/\text{kg}^2)(6 \times 10^{24}\ \text{kg})}{(6.4 \times 10^6\ \text{m} + 2 \times 10^5\ \text{m})}}$$

$$v = 7.8 \times 10^3\ \text{m/s}$$

194. (D) The two masses are designated $m = 10$ kg and $M = 30$ kg. The distance between them is 2 m. So, you can use Newton's law of universal gravitation to determine the gravitational force between them:

$$F_g = \frac{GmM}{r^2}$$

$$F_g = \frac{(6.7\times10^{-11}\ \mathrm{N\cdot m^2/kg^2})(10\ \mathrm{kg})(30\ \mathrm{kg})}{(2\ \mathrm{m})^2}$$

$$F_g = 5\times10^{-9}\ \mathrm{N}$$

195. (E) The car's mass is 1,000 kg, the centripetal force is 1.8×10^5 N, and the speed is 30 m/s. You can calculate the radius from the equation for centripetal force:

$$F_c = \frac{mv^2}{r}$$

$$r = \frac{mv^2}{F_c}$$

$$r = \frac{(1{,}000\ \mathrm{kg})(30\ \mathrm{m/s})^2}{(1.8\times10^5\ \mathrm{N})}$$

$$r = 5\ \mathrm{m}$$

196. (E) The skater's arm length (1 m) is the radius. She spins with a tangential velocity of 5 m/s. You can calculate the centripetal acceleration:

$$a_c = \frac{v^2}{r}$$

$$a_c = \frac{(5\ \mathrm{m/s})^2}{(1\ \mathrm{m})}$$

$$a_c = 25\ \mathrm{m/s^2}$$

197. (B) Mars's orbital distance is 2.3×10^{11} m. The mass of the Sun is 2×10^{30} kg and the mass of Mars is 6.4×10^{23} kg. Use Newton's law of universal gravitation to calculate the gravitational force that the Sun exerts on Mars:

$$F_g = \frac{Gm_1m_2}{r^2}$$

$$F_g = \frac{(6.7\times10^{-11}\ \mathrm{N\cdot m^2/kg^2})(6.4\times10^{23}\ \mathrm{kg})(2\times10^{30}\ \mathrm{kg})}{(2.3\times10^{11}\ \mathrm{m})^2}$$

$$F_g = 1.6\times10^{21}\ \mathrm{N}$$

198. (D) All the coins complete a circle of rotation in the same time period. Coin D traverses the greatest circumference, so it is traveling with the fastest tangential velocity.

199. (E) If the centripetal acceleration on the space shuttle remains the same, then the orbital radius will increase with the square of the velocity. So, if the velocity triples, then the orbital radius will increase by a factor of nine.

200. (D) Pick any two points on the graph and calculate the centripetal acceleration. You must square the velocity and divide it by the radius. You will find that all of them have the same centripetal acceleration, i.e., 10 m/s^2.

201. (D) The Moon's mass is 7.4×10^{22} kg and its radius is 1.7×10^6 m. The force of gravity experienced by a 70-kg astronaut standing on the lunar surface can be calculated with Newton's law of universal gravitation:

$$F_g = \frac{Gm_1m_2}{r^2}$$

$$F_g = \frac{(6.7\times10^{-11}\ \text{N}\cdot\text{m}^2/\text{kg}^2)(7.4\times10^{23}\ \text{kg})(70\ \text{kg})}{(1.7\times10^6\ \text{m})^2}$$

$$F_g = 120\ \text{N}$$

202. (B) The velocity of the wheel is its circumference divided by the period ($T = 1.6$ s). The wheel's radius is 0.5 m. You can then calculate the wheel's centripetal acceleration. Using the acceleration and the pebble's mass (10 g = 0.01 kg), you can calculate the centripetal force on the pebble:

$$v = \frac{2\pi r}{T}$$

$$v = \frac{2\pi(0.5\ \text{m})}{(1.6\ \text{s})}$$

$$v = 2\ \text{m/s}$$

$$a_c = \frac{v^2}{r}$$

$$a_c = \frac{(2\ \text{m/s})^2}{(0.5\ \text{m})}$$

$$a_c = 8\ \text{m/s}^2$$

$$F_c = ma_c$$

$$F_c = (0.01\ \text{kg})(8\ \text{m/s}^2)$$

$$F_c = 0.08\ \text{N}$$

203. **(B)** The Moon's mass is 7.4×10^{22} kg and its distance from the Earth is 3.8×10^{8} m. The Earth's mass is 6×10^{24} kg. You can calculate the gravitational potential energy of the Moon:

$$PE_g = \frac{Gm_1m_2}{r}$$

$$PE_g = \frac{(6.7\times10^{-11}\ \text{N}\cdot\text{m}^2/\text{kg}^2)(7.4\times10^{22}\ \text{kg})(6\times10^{24}\ \text{kg})}{(3.8\times10^{8}\ \text{m})}$$

$$PE_g = 7.8\times10^{28}\ \text{J}$$

204. **(D)** Saturn has an orbital period of 29 Earth years. Use Kepler's third law to determine its orbital distance:

$$T^2 = a^3$$

$$a = \sqrt[3]{T^2}$$

$$a = \sqrt[3]{(29\ \text{years})^2}$$

$$a = 10\ \text{AU}$$

205. **(B)** The coefficient of friction between the rubber tires of a car and dry concrete is $\mu = 0.64$. The radius of the turn is 10.0 m. You can calculate the maximum velocity of the car:

$$\mu = \frac{v^2}{gr}$$

$$v^2 = \mu gr$$

$$v = \sqrt{\mu gr}$$

$$v = \sqrt{(0.64)(10\ \text{m/s}^2)(10\ \text{m})}$$

$$v = 8\ \text{m/s}$$

206. **(D)** The radius of the top is 2 cm. The period of the top's spin is 0.06 s. Calculate the velocity first and then the centripetal acceleration:

$$v = \frac{2\pi r}{T}$$

$$v = \frac{2\pi(0.02\ \text{m})}{(0.06\ \text{s})}$$

$$v = 2.1\ \text{m/s}$$

$$a_c = \frac{v^2}{r}$$

$$a_c = \frac{(2.1\ \text{m/s})^2}{(0.02\ \text{m})}$$

$$a_c = 220\ \text{m/s}^2$$

207. (B) You know the mass (100 g = 0.1 kg), the radius of the circle (2 m), and the speed (12 m/s). You can calculate the centripetal force:

$$a_c = \frac{v^2}{r}$$

$$a_c = \frac{(12 \text{ m/s})^2}{(2 \text{ m})}$$

$$a_c = 72 \text{ m/s}^2$$

$$F_c = ma_c$$

$$F_c = (0.1 \text{ kg})(72 \text{ m/s}^2)$$

$$F_c = 7.2 \text{ N}$$

208. (B) If you apply Newton's law of universal gravitation to each planet and the equations for each planet have constant terms, G, and the mass of the star (M_s), you find the following data about relative gravitation:

Planet	Relative Mass	Relative Distance	Relative Gravitational Force ($GM_s m/r^2$)
A	2 m	r	2
B	m	0.1 r	100
C	0.5 m	2 r	1/8
D	4 m	3 r	4/9

The planet with the highest gravitational attraction is Planet B.

209. You know the mass of the satellite (1,000 kg), the altitude of the orbit (1,000 km), the Earth's mass (6.0×10^{24} kg), and the Earth's radius (6.4×10^6 m). Use this information in Newton's law of universal gravitation.

(a) The force of gravity on the satellite is the centripetal force on the satellite. The two are equal. The magnitude of the force of gravity is calculated as follows:

$$F_g = \frac{Gm_1m_2}{(R_E + r)^2}$$

$$F_g = \frac{(6.7\times10^{-11} \text{ N}\cdot\text{m}^2/\text{kg}^2)(1{,}000 \text{ kg})(6\times10^{24} \text{ kg})}{(6.4\times10^6 \text{ m} + 1.0\times10^6 \text{ m})^2}$$

$$F_g = 7300 \text{ N}$$

(b) Here is the magnitude of the satellite's tangential velocity:

$$v = \sqrt{\frac{GM_E}{r}}$$

$$v = \sqrt{\frac{(6.7\times10^{-11} \text{ N}\cdot\text{m}^2/\text{kg}^2)(6\times10^{24} \text{ kg})}{(6.4\times10^6 \text{ m} + 1\times10^6 \text{ m})}}$$

$$v = 7.4\times10^3 \text{ m/s}$$

(c) Calculate the gravitational potential energy of the satellite:

$$PE_g = \frac{Gm_1m_2}{(R_E + r)}$$

$$PE_g = \frac{(6.7\times10^{-11}\ \text{N}\cdot\text{m}^2/\text{kg}^2)(1000\ \text{kg})(6\times10^{24}\ \text{kg})}{(6.4\times10^6\ \text{m} + 1\times10^6\ \text{m})}$$

$$PE_g = 5.4\times10^{10}\ \text{J}$$

(d) Here is the value of the acceleration due to gravity (g') at this altitude:

$$g' = \frac{GM_E}{(R_E + r)^2}$$

$$g' = \frac{(6.7\times10^{-11}\ \text{N}\cdot\text{m}^2/\text{kg}^2)(6\times10^{24}\ \text{kg})}{(6.4\times10^6\ \text{m} + 1\times10^6\ \text{m})^2}$$

$$g' = 7.3\ \text{m/s}^2$$

210. (a) The free-body diagram of this situation looks like the following figure:

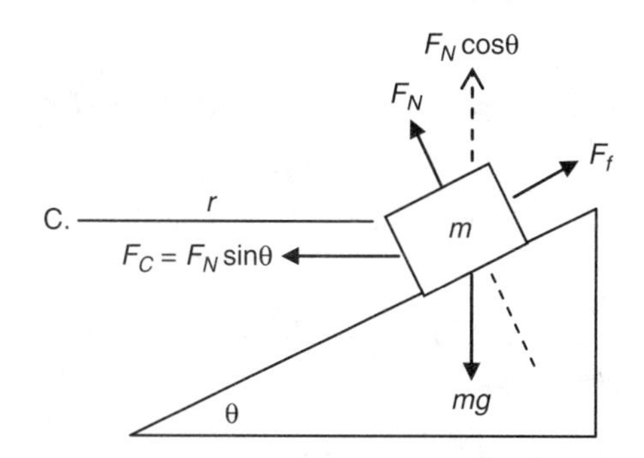

(b) We can calculate the car's maximum velocity by applying Newton's second law to the situation. The car's centripetal force is provided by the horizontal component of normal force ($F_c = F_N \sin\theta = mv^2/r$), which is provided by the road surface. The car does not move on or off the surface of the road, so the vertical component of the normal force must balance the weight of the car ($mg = F_N \cos\theta$). By dividing these two equations, we get an equation that we can solve for velocity:

$$F_c = F_N \sin\theta = \frac{mv^2}{r}$$

$$mg = F_N \cos\theta$$

$$\frac{F_N \sin\theta = \frac{mv^2}{r}}{F_N \cos\theta = mg}$$

$$\tan\theta = \frac{v^2}{rg}$$

$$v^2 = rg \tan\theta$$

$$v = \sqrt{rg \tan\theta}$$

$$v = \sqrt{(300 \text{ m})(10 \text{ m/s}^2)\tan 30^\circ}$$

$$v = 41 \text{ m/s}$$

(c) Calculate the centripetal force on the car:

$$F_c = \frac{mv^2}{r}$$

$$F_c = \frac{(1{,}000 \text{ kg})(41 \text{ m/s})^2}{(300 \text{ m})}$$

$$F_c = 5{,}600 \text{ N}$$

Chapter 8: Rotational Motion (For Physics C Students Only)

211. (B) R, θ, Z are the radial coordinates in a plane and the axis perpendicular to that plane.

212. (D) Rotational equivalent of mass depends on the distribution of mass and is equated to the resistance to change in angular velocity.

213. (B) Calculating its center of mass will show whether the star is rotating about its center of gravity or the center of mass of the system of one or more companions.

214. (E) The radial component of the Earth's acceleration adds a slight amount to the gravitational attraction everywhere but the poles.

215. (C) The planets' orbits are ellipses with the Sun at one focus.

216. (C) Mass is replaced by moment of inertia. All the equations are analogous, except that the moment of inertia replaces mass in rotational motion.

217. (D) Given that the moment of inertia of a slender rod is $ML^2/3$, the large mass of negligible size must be included to give the total moment of inertia of the system. So the total moment of inertia is then $mL^2/3 + ML^2$. Note that the masses of the individual parts of the system are (m) and (M), and the formula for the slender rod is a general formula.

218. (A) Like balancing a seesaw, the masses must be balanced about the center of mass. So $m \times R_1 = M \times R_2$ applies along with $L = R_1 + R_2$. Solving these two equations yields the correct answer.

219. (D) Angular momentum must be conserved in the system, so as the skater draws his arm in to his body, he is decreasing his moment of inertia. Thus his angular velocity must increase to conserve his angular momentum.

220. (B) Unlike degrees, which are arbitrarily determined, radians are based on the properties of a circle.

221. (D) The ball flies off tangent to the circle because the string is no longer holding the ball to the circular path; therefore, the ball has a velocity tangent to the circle. At the instant the string is cut, the velocity of the ball at that moment dictates how the ball will move.

222. (B) $I = m \times R^2$. All of the mass elements are the same distance from the axis R.

$$I = \int R^2 dm$$

$$I = R^2 \int dm$$

$$I = mR^2$$

223. (D) Although some of the other units are units of angular rotation, radians/s is the generally accepted unit.

224. (E) 7,770 m/s is given by the formula, $v = R \times \omega$, in which v is the orbital velocity, R is the radius of the junk's orbit, and ω is the angular rotation of the junk or $2\pi/(90 \times 60)$ radians/s.

225. (E) Kinetic energy of the orbiting body is dependent on the mass, its speed, and its moment of inertia. The kinetic energy = one-half $\times m \times v^2$ + one-half $\times I \times \omega^2$. In this case, $I = mR^2$ because the mass is small and can be considered a point mass.

226. (C) A torque is a vector because it is the cross product of a force vector and a position vector.

227. (D) We want the Earth and space platforms to be moving with the same velocity. Velocity is $R \times \omega$ and set V_{Earth} to V_{space}. Solving this equation indicates that the platform should be moving at 6 radians/day, so it will remain above the Earth platform moving at 2π radians/day.

228. (C) Newton's first law has six equations: the three for summation of forces along the three axes and the three summations of moments about the three axes.

229. (E) Using the right-hand rule, the force vector creating the torque is crossed into the position vector of the force, which gives the torque's magnitude and direction.

230. (C) When a body is moving in circular motion, the types of acceleration acting on it are radial acceleration, tangential acceleration, and angular acceleration.

231. (D) The velocity for this case is given by $v = \sqrt{\mu G R}$ in which v is the speed, μ is the coefficient of friction, G is the acceleration of gravity, and R is the radius of the curve. Note that the speed does not depend on the mass of the car.

232. (C) The bead's position is a function of the speed of revolution. No matter how fast the wire spins, the bead will not get half way up the wire. The bead can move to positions below that point if the speed is fast enough. If the speed is slow enough, the bead will slide to the bottom of the axis of rotation.

233. (D) A formula for speed in circular motion is $v = 2 \times \pi \times R / T$ in which v is the speed, R is the radius of the circle, and T is the time taken to make one circuit of the circle. Inserting all the givens into this formula yields the answer.

234. (B) At the bottom of the loop it will be moving the fastest and at the top the slowest. Gravity slows the car at the top of the loop and speeds the car to its fastest speed at the bottom of the loop.

235. (E) Gravity and radial acceleration act on the rider's weight to produce the stated effect.

236. (C) $v = \sqrt{2gh / \left(1 + \frac{M}{2m}\right)}$ is obtained by conservation of energy when considering the movement of the winch, the height of the anchor above the ocean floor, and the falling speed of the anchor.

237. (C) $MR^2/2 + MR^2$ uses the formula of a solid cylinder around its central axis, $MR^2/2$. The parallel–axis theorem then adds MR^2 to account for the new axis.

238. (D) 6.16×10^6 W is the result when using the power and torque formula, $P = \omega \times \tau$, in which P is power, ω is the angular velocity, and τ is the torque produced by the blades.

239. In this problem, angular momentum is conserved, so the momentum of the BBs is $m \times v \times d$, where m is the mass of the BBs, v is the velocity, and d is the moment arm. This gives 0.001 kg m^2/s. The target's momentum after the hit is $I \times \omega$, in which I is the target's moment of inertia and ω is the moment of inertia to be found. So $\omega = 0.001/0.015$ rad/s or 0.067rad/s.

240. In this problem, the normal force of the car against the road and the weight of the car are the forces acting on the car. The car also has a radial acceleration toward the center of the curve. Newton's second law applies to this question, so $\Sigma \text{F}_x = m \times a_{\text{rad}}$ and $\Sigma \text{F}_y = 0$. The normal force (N) can be broken into its x and y components based on the bank angle, B. So Newton's equations become: $\text{N} \times \sin B = m \times a_{\text{rad}}$ and $\text{N} \times \cos B + (-m \times g) = 0$. Solving for B, we find $B = \arctan a_{\text{rad}}/g$ or $v^2/(R \times g)$ in which v is the car's speed and R is the radius of the curve. In this case, $B = 49°$.

Chapter 9: Simple Harmonic Motion

241. (D) The spring's opposing force is the vector along the incline, Mg*cosine θ. Then, from the mass/spring period formula, $T = 2\pi\sqrt{M \cos\theta / K}$ is the period of oscillation.

242. (E) The restoring force is due to the inertia of the small mass or mass ratio. The rod has a rotational spring constant of C. Then, similar to a coiled spring, the period of oscillation is $T = 2\pi \times \sqrt{M \times R \div C}$.

243. (D) The pendulum swings and eventually causes the spring to oscillate. This cycle repeats in a frictionless system.

244. (B) The period of rotation is 27.3 days. Since the frequency is the inverse of the period, the period is $F = 1 \div 27.3$ days or 0.037 revolutions/day.

245. (C) Because the bell and clapper have the same period, they move as one when rung.

246. (D) The period of a pendulum depends on its length. So, by making the clapper longer, its period would be longer and out of phase with the bell. Therefore, the bell will ring with a longer clapper.

247. (E) By raising your center of gravity, you add potential energy to the system, and this causes the swing to go higher. By kicking out to the top of the swing, kinetic energy is added to the system.

248. (D) It swings in one plane. As the Earth moves under the pendulum, the pendulum demonstrates that the Earth rotates.

249. (D) It will swing until its potential energy is equal to the original potential energy, i.e., the horizontal or 180°.

250. (C) The kinetic energy in the raised ball is transferred through the balls to the only ball that can move. Thus, it rises.

251. (B) Because the pendulum's mass moves in response to gravity, the restoring force is a function of the mass and gravity.

252. (E) The acceleration of the ship provides an artificial gravity as long as it is accelerating.

253. (A) Yes, it is moving periodically.

254. (E) It will move in a straight line tangential to where it was released and curve downward because of gravity.

255. (C) The water tank is 180° out of phase with the ship's hull so the rolling of the ship is cancelled by the rolling of the water in the tank.

256. (B) Using the formula for the period of a pendulum, $2\pi\sqrt{L \div g}$, the period is 16.43 seconds and the frequency is the inverse of the period or 0.06 cycles/s.

257. (D) is found by using the formula for the period of a mass spring system, $= 2\pi \times \sqrt{M \div K}$, and by using the formula for two springs in parallel, $k_{eg} = k_1 + k_2$.

258. (B) is found by using the formula for the period of a mass spring system, $= 2\pi \times \sqrt{M \div K}$, and by using the formula for two springs in series, $1/k_{eg} = 1/k_1 + 1/k_2$.

259. (D) is found by using the formula for the period of a mass spring system, $= 2\pi \times \sqrt{M \div K}$, and by using the formula for two springs in parallel, $k_{eg} = k_1 + k_2$.

260. (B) is found by using the formula for the period of a mass spring system, $= 2\pi \times \sqrt{M \div K}$, and by using the formula for two springs in series, $1/k_{eg} = 1/k_1 + 1/k_2$.

261. (D) 4.4 s for both cases. It is found by using the formula for the period of a mass spring system, $= 2\pi \times \sqrt{M \div K}$, and knowing that both cases are frictionless.

262. (A) Determine the answer by using the formula for the period of a mass and spring system, $= 2\pi \times \sqrt{M \div K}$. You also know that the three springs are parallel, so $k_{eq} = k_1 + k_2 + k_3$.

263. (D) The spring constant determines the stretch of the spring, $= M \times g/k$, and the period of a mass spring system, $T = 2\pi \times \sqrt{M \div K}$.

264. (A) The springs are supporting the compressor in parallel. Then, use the formula of frequency, $f = (1 \div (2 \times \pi)) \times \sqrt{.03/(8 \times 9.8)}$. (Remember to convert the mass to weight.)

265. (D) Using the formula for the period of a pendulum, $T = 2\pi\sqrt{L \div g}$, solve for L given that $T = 3$.

266. (B) To answer this question, you must know the speed of light, which is 299,800 m/s. From the frequency formula, we get the period of the vibration, which is $1/6.662 \times 10^{-11}$ or 1.502×10^{-12} s. This is the time it takes the light to move 1 wavelength. Multiply this period by the speed of light.

267. From the falling body equations, $-x_0 = v_0 \times t + 0.5 \times a \times t^2$, where x_0 is zero, v_0 is zero, a is the acceleration of gravity, and t is one-half of the period of the periodic motion. Because the collision of the ball with the plate is elastic, the full energy of the dropped ball is returned to the ball. So inserting $t = 1$ into the falling body equation and solving for x we get $x = 4.9$ m.

268. (D) This will allow the pendulum to oscillate and not fall over because the plate is providing a restoring force to the pendulum's bob, which counteracts gravity.

269. Yes, it will be periodic because of conservation of energy. The ball will drop down the hole until it reaches its maximum velocity at the sphere's center. So, its original potential energy at the surface is equal to its kinetic energy at the center. Then, its kinetic energy will carry it to the surface where its velocity will be zero and the gravitational pull of the sphere with cause the ball to fall back into the hole and so on. The ball drops 6,400 km, or 6.4 million m, to the center of the sphere. From the falling-body equations, $-x_0 = v_0 \times t + 0.5 \times a \times t^2$, where x_0 is zero, v_0 is zero, a is the acceleration of gravity, and t is one-half of the period of the periodic motion. Solving this gives $T = 2586$ s or 38 min.

270. The maximum and minimum amplitudes will occur when ωt is $\pi/2$ and $3\pi/2$. The time constant is $\omega = \sqrt{k/m}$, which is also related to the period and frequency. Therefore, $\omega = 0.0005$ radians/s. Using this in the period formula, $T = 281$ s, and the inverse of the period is the frequency, or $f = .004$ Hz. The maximum and minimum are equal to 20 cos $\pi/2$ and 20 cos $3\pi/2$ or 20 cm and −20 cm.

Chapter 10: Thermodynamics

271. **(D)** The initial length of the railroad section is 1.0 m. The change in temperature is 15°C. The coefficient of thermal linear expansion for iron is 1.2×10^{-5}°C^{-1}. So:

$$\Delta L = \alpha L_0 \Delta T$$

$$\Delta L = (1.2\times10^{-5}\,^\circ\mathrm{C}^{-1})(1.0\ \mathrm{m})(15^\circ\mathrm{C})$$

$$\Delta L = 1.8\times10^{-4}\ \mathrm{m} = 0.18\ \mathrm{mm}$$

272. **(B)** You have 20 moles of a gas. The volume of the container is 1 m^3 container. The temperature of the gas is 125°C. Use the ideal gas law to determine the pressure. Remember that 1 atm = 1×10^5 N/m^2.

$$PV = nRT$$

$$P = \frac{nRT}{V}$$

$$P = \frac{(20\ \mathrm{moles})(8\ \mathrm{J/mol\cdot K})(400\ \mathrm{K})}{(1.0\ \mathrm{m}^3)}$$

$$P = 6.4\times10^4\ \mathrm{N/m}^2$$

$$P = 0.6\ \mathrm{atm}$$

273. **(C)** You know the number of moles of gas (2), the temperature (25°C = 300 K), and the universal gas constant (8 J/mol · K). You can calculate the internal energy of the gas (U):

$$U = \frac{3}{2}nRT$$

$$U = \frac{3(2\ \mathrm{moles})(8\ \mathrm{J/mol\cdot K})(300\ \mathrm{K})}{2}$$

$$U = 7200\ \mathrm{J}$$

274. **(D)** You know the mass of an oxygen atom (5×10^{-26} kg), the temperature (25°C = 300 K), and the Boltzmann constant (1.4×10^{-23} J/K). You can find the average speed (v_{rms}) of an oxygen molecule:

$$v_{rms} = \sqrt{\frac{3k_bT}{m}}$$

$$v_{rms} = \sqrt{\frac{3(1.4\times10^{-23}\ \mathrm{J/K})(300\ \mathrm{K})}{(5\times10^{-26}\ \mathrm{kg})}}$$

$$v_{rms} = 500\ \mathrm{m/s}$$

275. (B) In an adiabatic process, 0.5 moles of gas at 1,000 K expands to reach a final temperature of 500 K. You know that the process was adiabatic, so $Q = 0$. You also know the gas expanded and the temperature dropped, so work was done by the gas and the work value should be negative. You can calculate the work from the first law of thermodynamics:

$$\Delta U = Q + W$$

$$Q = 0$$

$$\Delta U = W$$

$$\Delta U = \frac{3}{2}nR(\Delta T)$$

$$W = \frac{3}{2}nR(\Delta T)$$

$$W = \frac{3}{2}(0.5 \text{ moles})(8 \text{ J/mol} \cdot \text{K})(500 \text{ K} - 1000 \text{ K})$$

$$W = -3000 \text{ J}$$

276. (C) The heat source of the engine is 5,000 J of thermal energy. The work done by the engine is 2,500 J. You can calculate the efficiency of the engine:

$$e = \frac{W}{Q_H}$$

$$e = \frac{(2{,}500 \text{ J})}{(5{,}000 \text{ J})}$$

$$e = 0.5 \text{ or } 50\%$$

277. (C) An ideal heat engine has an efficiency of 20%. The heart reservoir has a temperature of 200°C. You can calculate the temperature of the heat sink from the ideal heat engine equation:

$$e_{ideal} = \frac{T_H - T_C}{T_H}$$

$$T_H - T_C = e_{ideal} T_H$$

$$T_C = T_H - e_{ideal} T_H$$

$$T_C = T_H (1 - e_{ideal})$$

$$T_C = (475 \text{ K})(1 - 0.2)$$

$$T_C = 380 \text{ K or } 110^{\circ}\text{C}$$

278. (B) Work is 500 J and atmospheric pressure is constant at 1×10^5 N/m^2. Because work was done on the gas, the value of work must be positive. Also, because pressure was constant, the change in volume must be negative:

$$W = -p\Delta V$$

$$\Delta V = -\frac{W}{p}$$

$$\Delta V = -\frac{(500 \text{ J})}{(1 \times 10^5 \text{ N/m}^2)}$$

$$\Delta V = -0.005 \text{ m}^3$$

279. (A) On a pressure–volume diagram, the work done is the area under the curve; in this case, it is a trapezoid. The work was done by the gas as stated in the question, so the sign of the work must be negative:

$$A = \frac{1}{2}(b_1 + b_2)h$$

$$A = \frac{1}{2}[(1 \times 10^5 \text{ N/m}^2) + (5 \times 10^5 \text{ N/m}^2)](4 \times 10^{-3} \text{ m}^3)$$

$$A = 1{,}200 \text{ J}$$

$$W = -1{,}200 \text{ J}$$

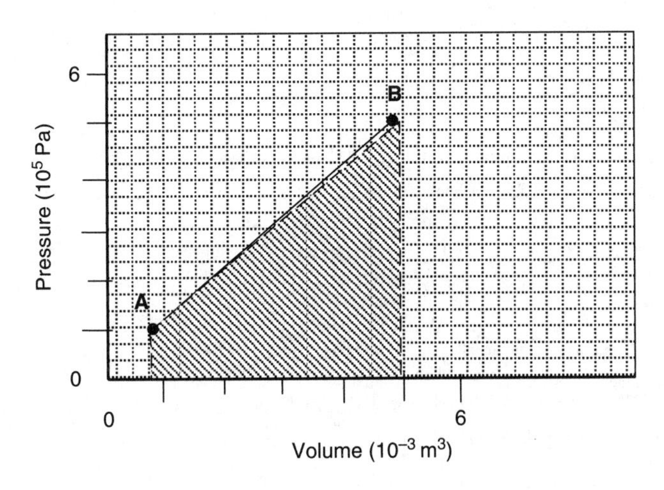

280. (D) In an isochoric process, there is no change in volume. Therefore, there is no work done on or by the gas.

281. (A) A copper rod has a diameter of 2 cm and a length of 0.5 m (the area is 3×10^{-4} m^2). The rod's temperature increases from 25°C to 50°C. The thermal conductivity of copper is 390 J/s·m·°C. You can calculate the rate of heat transfer into the copper rod:

$$H = \frac{kA\Delta T}{L}$$

$$H = \frac{(390 \text{ J/s}\cdot\text{m}\cdot{}^\circ\text{C})(3\times10^{-4}\text{ m}^2)(25^\circ\text{C})}{(0.5\text{ m})}$$

$$H = 6 \text{ J/s}$$

282. (A) The mass of the crate is 20 kg, the initial velocity is 0.5 m/s, the final velocity is 0 m/s, and the time is 25 s. Calculate the change in energy as the block moves, then use this value to calculate the rate of energy transfer or power:

$$KE = \frac{1}{2}m(v_f^2 - v_i^2)$$

$$KE = \frac{1}{2}(20\text{ kg})[(0.5\text{ m/s})^2 - (0\text{ m/s})^2]$$

$$KE = 2.5\text{ J}$$

$$P = \frac{\Delta KE}{t}$$

$$P = \frac{(2.5\text{ J})}{(25\text{ s})}$$

$$P = 0.1\text{ W}$$

283. (D) One kJ of thermal energy (1,000 J) is transferred to a gas in a cylinder with a movable piston. At the same time, 200 J of work is done on the system. You can calculate the change in internal energy of the system from the first law of thermodynamics:

$$\Delta U = Q + W$$

$$\Delta U = (1{,}000\text{ J}) + (200\text{ J})$$

$$\Delta U = 1{,}200\text{ J}$$

284. (C) In an isothermal process, ΔU is zero. So, if thermal energy is added to the gas, then an equal amount of work must be done by the gas:

$$\Delta U = Q + W$$

$$0 = Q + W$$

$$W = -Q$$

$$W = -(500\text{ J})$$

$W = -500$ J, The gas does work as indicated by the negative sign.

285. (B) Use the following equation to determine the material that the rod is made of:

$$\Delta L = \alpha L_0 \Delta T$$

$$\alpha = \frac{\Delta L}{L_0 \Delta T}$$

$$\alpha = \frac{(1.7 \times 10^{-4}\ \text{m})}{(1.0\ \text{m})(10^\circ\text{C})}$$

$$\alpha = 1.7 \times 10^{-5}\ {}^\circ\text{C}^{-1}$$

From the table, the value matches copper.

286. (D) Two moles of oxygen gas are in a container with a movable piston. The temperature is 25°C. The pressure is 6,400 Pa. Use the ideal gas equation to find the volume of the gas:

$$PV = nRT$$

$$V = \frac{nRT}{P}$$

$$V = \frac{(2.0\ \text{moles})(8\ \text{J/mol} \cdot \text{K})(400\ \text{K})}{(6{,}400\ \text{N/m}^2)}$$

$$V = 1.0\ \text{m}^3$$

287. (E) You have 1 mole of hydrogen gas with an internal energy of 4.3 kJ (4,300 J). You can calculate the temperature of the gas:

$$U = \frac{3}{2} nRT$$

$$2U = 3nRT$$

$$T = \frac{2U}{3nR}$$

$$T = \frac{2(4{,}300\ \text{J})}{3(1\ \text{mole})(8\ \text{J/mol} \cdot \text{K})}$$

$$T = 360\ \text{K}$$

288. (D) The engine has a 20% efficiency rating and the input thermal energy is 2 kJ (2,000 J). Use the efficiency equation to calculate the work done by the engine:

$$e = \frac{W}{Q_H}$$

$$W = eQ_H$$

$$W = (0.2)(2{,}000\ \text{J})$$

$$W = 400\ \text{J}$$

289. (E) The heat source for an ideal heat engine has a temperature of 1,000°C (1,273 K). The temperature of the heat sink is 100°C (373 K). Calculate the engine's efficiency with the ideal heat engine equation:

$$e_{ideal} = \frac{T_H - T_C}{T_H}$$

$$e_{ideal} = \frac{(1{,}273\text{ K}) - (373\text{ K})}{(1{,}273\text{ K})}$$

$$e_{ideal} = 0.7 \text{ or } 70\%$$

290. (A) You change the temperature of one mole of nitrogen gas from 25°C to −175°C. You can calculate the change of the gas' internal energy. The change in energy must be negative since the temperature decreases:

$$\Delta U = \frac{3}{2} nR\Delta T$$

$$\Delta U = \frac{3(1\text{ mole})(8\text{ J/mol}\cdot\text{K})(-200\text{ K})}{2}$$

$$\Delta U = -2{,}400\text{ J}$$

291. (A) The temperature changes from 20°C to 40°C. The coefficient of thermal linear expansion for concrete is 1.2×10^{-5} °C^{-1}. You can find the percentage change in the length of the slab:

$$\Delta L = \alpha L_0 \Delta T$$

$$\frac{\Delta L}{L_0} = \alpha \Delta T$$

$$\frac{\Delta L}{L_0} = (1.2 \times 10^{-5}\,^\circ\text{C}^{-1})(20^\circ\text{C})$$

$$\frac{\Delta L}{L_0} = 2.4 \times 10^{-4} = 0.024\%$$

292. (C) The volume of the gas increases threefold when at a constant pressure. According to $W = -p\Delta V$, the sign of the work must be negative. This indicates the gas does work at a threefold increase.

293. (B) The volume of the cylinder is 0.001 m^3, the pressure is 2,500 N/m^2, and the temperature is 25°C (300 K). You can calculate the number of molecules from the Boltzmann version of the ideal gas law:

$$PV = Nk_bT$$

$$N = \frac{PV}{k_bT}$$

$$N = \frac{(2{,}500\ \text{N/m}^2)(0.001\ \text{m}^3)}{(1.4\times10^{-23}\ \text{J/K})(300\ \text{K})}$$

$$N = 6\times10^{20}$$

294. (E) Use the following equation to solve the question:

$$H = \frac{kA\Delta T}{L}$$

$$HL = kA\Delta T$$

$$\Delta T = \frac{HL}{kA}$$

$$\Delta T = \frac{(1\ \text{J/s})(1.0\ \text{m})}{(110\ \text{J/s}\cdot\text{m}\cdot{}^\circ\text{C})(3\times10^{-4}\ \text{m}^2)}$$

$$\Delta T = 30^\circ\text{C}$$

295. (E) The internal energy change during an isochoric process is 1,000 J. In an isochoric process, work is zero. Calculate how much heat was transferred using the first law of thermodynamics:

$$\Delta U = Q + W$$

$$W = 0$$

$$\Delta U = Q$$

$$W = 1{,}000\ \text{J}$$

$W = -500$ J, the gas does work as indicated by the negative sign.

296. (E) The temperature of steam is 125°C (400 K). You can calculate the mass of a water molecule (3×10^{-26} kg) from the formula weight of water and Avogadro's number:

$$v_{rms} = \sqrt{\frac{3k_bT}{m}}$$

$$v_{rms} = \sqrt{\frac{3(1.4\times10^{-23}\ \text{J/K})(400\ \text{K})}{(3\times10^{-26}\ \text{kg})}}$$

$$v_{rms} = 750\ \text{m/s}$$

297. **(E)** The force is 2,000 N, the piston moves 1 m, and the change occurs adiabatically ($Q = 0$). You can calculate the change in internal energy of the gas by using the first law of thermodynamics:

$$\Delta U = Q + W$$
$$Q = 0$$
$$\Delta U = W$$
$$W = Fd$$
$$W = (2{,}000\text{ N})(1\text{ m})$$
$$\Delta U = W = 2{,}000\text{ J}$$

298. **(C)** An air conditioner must have work done to remove thermal energy from a cold sink to a hot reservoir.

299. (a) Work is the area under the curve (indicated on the graph).

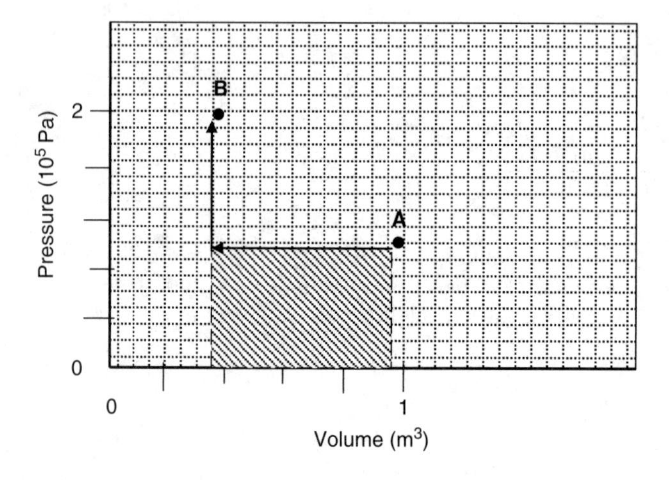

(b) First arrow indicates an isobaric compression:

$$W = -p\Delta V$$
$$W = -(1\times10^5\text{ N/m}^2)(0.4\text{ m}^3 - 1.0\text{ m}^3)$$
$$W = -6\times10^4\text{ J}$$

Second arrow indicates isochoric change in pressure

$$\therefore\ W = 0$$

The total work is -6×10^4 J.

(c) $\Delta U = Q + W$

$Q = 0$ for adiabatic process

$\Delta U = W$

$\Delta U = -6 \times 10^4 \text{ J}$

(d) $\Delta U = \frac{3}{2} nR\Delta T$

$2\Delta U = 3nR\Delta T$

$\Delta T = \frac{2\Delta U}{3nR}$

$\Delta T = \frac{2(-6 \times 10^4 \text{ J})}{3(10 \text{ moles})(8 \text{ J/mol} \cdot \text{K})}$

$\Delta T = 500 \text{ K}$

(e) $\Delta T = T_f - T_i$

$T_f = T_i + \Delta T$

$T_f = (50 \text{ K}) + (500 \text{ K})$

$T_f = 550 \text{ K}$

300. (a) $\Delta L = \alpha L_0 \Delta T$

$\Delta L = (1.9 \times 10^{-5} \,^\circ\text{C}^{-1})(0.1 \text{ m})(200^\circ\text{C})$

$\Delta L = 3.8 \times 10^{-4} \text{ m} = 0.38 \text{ mm}$

(b) $\% \text{ change} = \frac{\Delta L}{L_0} \times 100$

$\% \text{ change} = \frac{(3.8 \times 10^{-4} \text{ m})}{(0.1 \text{ m})} \times 100$

$\% \text{ change} = 0.38 \%$

(c) $Q = \frac{kA\Delta Tt}{L}$

$Q = \frac{(420 \text{ J/s} \cdot \text{m} \cdot {}^\circ\text{C})(7.85 \times 10^{-7} \text{ m}^2)(200^\circ\text{C})(30 \text{ s})}{(0.1 \text{ m})}$

$Q = 19.8 \text{ J}$

Chapter 11: Fluid Mechanics

301. (B) Using the formula of pressure = force/area, the force is 1 N, and the area is $\pi \times (0.02\text{ m})^2/4$. So, the pressure in the syringe is 3,180 Pa (N/m^2).

302. (C) 270,000 Pa is found from the static pressure head formula, $P = \delta \times G \times h$. P is the static pressure, δ is the density of water (or 1,000 kg/m^3), G is the acceleration of gravity (9.8 m/sec^2), and h is the height of the water column from the top of the water level to the point of outlet, or 27.5 m.

303. (D) 620 Pa is found from the pressure formula, $P = F/A$, where F is the weight of the manhole cover or 50 kg $\times$ 9.8 N m/s^2, and A is the area of the manhole cover or 1 $\text{m}^2 \times \pi/4$.

304. (D) The glass will overflow because the ice floats but displaces a volume of water to float. Since the glass is full, adding ice will overflow water from the glass.

305. (C) When Torricelli took the barometer into the mountains, the lower air density at those heights caused the mercury column to decrease. When he returned to Faenza, a low-pressure front with its usual rains had moved into Faenza. This caused the barometer to read the same as it did in the mountains.

306. (C) 18,600 N is found by multiplying the cross-sectional area of the sphere by the difference between the inside and outside air pressures. Assuming standard air pressure of 101,325 Pa, then $\pi \times .255^2 \times (101{,}325 - 10{,}132.5) = 18{,}600$ N.

307. (B) When the rock is in the boat, the water level will increase because the weight of the boat and rock equals the weight of the displaced water. When the rock is dumped into the water, only the volume of the rock is displaced. Because the rock is denser than water, the water level in the pool will fall because it is only supporting the boat.

308. (A) 50,400,000 N is obtained from Archimedes' principle wherein the buoyant force is equal to the liquid's density multiplied by the volume submerged and the acceleration of gravity. The buoyant force must be counteracted by adding a weight that is equal to or greater than the buoyant force.

309. (D) 1.3 m is calculated using Archimedes' principle. The salvage ship displaces 75 m^3/0.5 m on the load line. So 75,600 N (volume submerged $\times$ density of sea water $\times$ acceleration of gravity) are displaced for each 0.5 m on the load line. The total load of caryatids is 200,000 N. Dividing the load by the displacement weight tells us that 2.6 displacement weights are needed to support the caryatids. On the load line this equals 2.6 $\times$ 0.5 m, or the ship rides 1.3 m lower in the water with the caryatids on board.

310. (B) A pitot tube is an instrument used to measure flow. Because of continuity, if there is 50,000 L/s at the outlet, there must be 50,000 L/s flowing anywhere in the pipeline. The velocity of the flow is affected by the diameter of the pipe. So, 50,000 L/s = 50 m^3/s. Divide this by the areas of the three points in the pipe: 7.07, 3.14, 38.48 m^2.

311. (D) The pump moves 5 cc of oil with each stroke. The jack cylinder is 3 cm in diameter with an area of 7 cm^2. Dividing 5 cc by 7 cm^2 gives the height raised with each stroke.

312. (B) 270,000 N divided by 2 m^2 is 140,000 Pa.

313. (D) Use Torricelli's law: $v = h$, where h is 2.75 m, and g is 9.8 m/s^2. Then, the velocity is 7.3 m/s.

314. (D) Torricelli's law can also be expressed as $v = \sqrt{2P/\delta}$. Using the velocity from the previous problem, 7.3 m/s^2, and the water's density, 1,000 kg/m^3, we solve for P, which turns out to be 27,000 Pa. The force needed to cap the outlet will be $F = P/A$, where A is the outlet's area. So, the force needed to cap the pipe is 6,700,000 N, which could not be applied by a man.

315. (C) Using $v = \sqrt{2P/\delta}$ again, P is the maintained pressure, 690 kPa, and δ is the density of water, 1,000 kg/m^3. Then, v is 40 m/s.

316. (B) For every 1,000 m, the water drops 0.3 m. So, 15×0.3 gives a pressure head of 4.5 m. Then, using the two formulas of Torricelli's law, $v = \sqrt{2P/\delta}$ and $v = \sqrt{2gh}$, we can solve for P, which is 0.04 Pa.

317. Standard air pressure is 101,325 N/m^2. So, the height of each column of liquid is balanced by the height of liquid in the tube. $P_{atm} = h \times \delta$. The column heights are 10.33 m, 13.1 m, and 5.642 m, respectively.

318. (B) First, find the volume of each object from the equation, $V_{obj} = W_{obj} \div \delta_{obj}$. Then, the buoyant force is the weight of the displaced water: $F = V_{obj} \times \delta_{water}$.

319. (D) In the same manner as **Question 318**, the buoyant force is 20 N × 9,807 N/m^3/70,600 N/m^3 or 2.8 N. Then, the amount of force needed to pull the skillet up is 20 N − 2.8 N = 17.2 N.

320. (E) Once running, there is more liquid on the downstream side than on the upstream side. The extra weight helps pull the liquid through the siphon. Some liquids can be siphoned in a vacuum.

321. (E) Once the pressure is reduced enough to allow bubbles of the liquid's vapor to form in the siphon, then the stream is broken and it will stop.

322. (C) Water from the tank causes the water level to be above the toilet's siphon bend and starts siphoning.

323. (D) Use the Torricelli and Bernoulli equations, $h = P \div (\delta \times g)$, where h is the height the stream can reach, P is the pressure in the hose, δ is the mass density of the water, and g is the acceleration of gravity.

324. (E) Gravitational effects are negligible, but water vapors change throughout the atmosphere.

325. **(E)** The wind is funneled between the buildings and the flow is continuous, but Bernoulli's equation of conservation of energy explains why the wind speeds up between the buildings.

326. **(C)** Pressure on the higher side forces the mercury up to a level at which the column's weight equals the pressure difference. Convert 200 mm to 0.2 m. The pressure difference is then 0.2 m × 133,000 N/m^3, which equals 27,000 N/m^2 or 27,000 Pa.

327. **(D)** This can be calculated when the various accelerations are known.

328. **(C)** The pressure varies from 1 m to 1.25 m. So, the average depth of the porthole is 1.12 m. The pressure at this depth is $h \times \delta \times g$, or 11,000 Pa. The porthole should hold under this small pressure.

329. The frequency is $\frac{1}{2\pi} \times \sqrt{\frac{\delta g A}{m}}$, where δ is the mass density of the water, g is the acceleration of gravity, A is the cross section of the log, and m is the mass of the log. Effectively, the log is supported by a spring that is the buoyant force on the log, or $F_b = \delta g A h$. In this case, Ah is the submerged volume of the log. Rewrite the buoyant force formula, $F_b = (\delta g A)h$, in which $(\delta g A)$ is the spring constant. Then, use the formula for frequency: $f = \frac{1}{2\pi} \times \sqrt{\frac{\delta g A}{m}}$.

330. The liquid in a U tube oscillates around the mean level when the height of the liquid is the same on both sides. So, the displacement about the mean in a U tube is l/2 where l is the height difference used to start the oscillation. Then, use the formula for the frequency of a pendulum, $f = \frac{1}{2\pi} \times \sqrt{\frac{2g}{l}}$.

Chapter 12: Electrostatics

331. **(D)** The electric charge of a proton is 1.6×10^{-19} C. The object has a charge of 8.0×10^{-19} C. So, it has five more protons than electrons.

332. **(C)** Only charged particles are affected by electric fields. Protons travel in the direction of the field, while neutrons travel opposite the direction of the field. So, the electron will be deflected to the left, the proton will be deflected to the right, and the neutron will pass through the field.

333. **(D)** The charge on an electron is -1.0×10^{-19} C and the strength of the electric field is 100 N/C. So, the force on the electron can be calculated:

$$F = qE$$

$$F = (-1.60 \times 10^{-19}\ \text{C})(100\ \text{N/C})$$

$$F = -1.60 \times 10^{-19}\ \text{C}$$

The negative sign indicates that the electron moves opposite the direction of the field (i.e., to the left).

334. (D) An electron will be attracted toward the positive charge, so it takes work to move the electron from the 20-V equipotential line to the 40-V equipotential line. The work is the difference between the electrical potential at the two lines. The distance is irrelevant.

$$W = \Delta PE = q\Delta V = q(V_2 - V_1)$$

$$W = (1.6\times10^{-19}\text{ C})(40\text{ V} - 20\text{ V})$$

$$W = 3.2\times10^{-19}\text{ J}$$

335. (D) The charge on a proton is 1.6×10^{-19} C. The distance between them is 1×10^{-6} m. The protons have the same charge, so the force will be repulsive. Use Coulomb's law to find the magnitude of the force:

$$F = \frac{kq_1q_2}{r^2}$$

$$F = \frac{(9\times10^9\text{ N}\cdot\text{m}^2/\text{C}^2)(1.6\times10^{-19}\text{ C})(1.6\times10^{-19}\text{ C})}{(1.0\times10^{-6}\text{ m})^2}$$

$$F = 2.3\times10^{-16}\text{ N}$$

336. (E) The change in electrical potential energy ($\Delta EPE = PE_B - PE_A$) is -5.0×10^{-5} J, and the magnitude of the test charge is 1 μC (1×10^{-6} C). You can calculate the voltage difference between Points A and B. Because the ΔEPE is negative, Point A must have a higher electrical potential energy than Point B:

$$\Delta EPE = q\Delta V$$

$$\Delta V = \frac{\Delta EPE}{q}$$

$$\Delta V = \frac{(-5\times10^{-5}\text{ J})}{(+1\times10^{-6}\text{ C})}$$

$$\Delta V = -50\text{ V}$$

337. (D) Test Charge D is the closest to the positive point charge, so it has the greatest force acting on it.

338. (E) Charge E is the farthest away from the positive charge, so it has the greatest electrical potential energy.

339. (B) Charges A and C occupy the same equipotential line, so they have the same electrical potential energy.

340. (E) Moving Charge D to Charge E requires the most work because the force of attraction at D is the greatest.

341. **(E)** The separation between the parallel plates is 1 mm or 1×10^{-3} m. The voltage across the plates is 9.0 V. You can calculate the magnitude of the electric field:

$$E = \frac{V}{d}$$

$$E = \frac{(9.0 \text{ V})}{(1 \times 10^{-3} \text{ m})}$$

$$E = 9{,}000 \text{ V/m}$$

342. **(A)** The mass of the crate is 20 kg, the initial velocity is 0.5 m/s, the final velocity is 0 m/s, and the time is 25 s. Calculate the change in energy as the block moves, and use this value to calculate the rate of energy transfer or power:

$$KE = \frac{1}{2}m(v_f^2 - v_i^2)$$

$$KE = \frac{1}{2}(20 \text{ kg})[(0.5 \text{ m/s})^2 - (0 \text{ m/s})^2]$$

$$KE = 2.5 \text{ J}$$

$$P = \frac{\Delta KE}{t}$$

$$P = \frac{(2.5 \text{ J})}{(25 \text{ s})}$$

$$P = 0.1 \text{ W}$$

343. **(B)** The capacitance is 1 μF or 1×10^{-6} F. The voltage is 12.0 V. You can calculate the charge as follows:

$$Q = CV$$

$$Q = (1 \times 10^{-6} \text{ F})(12 \text{ V})$$

$$Q = 1 \times 10^{-5} \text{ C}$$

344. **(A)** The area of the plates is 1 cm^2 (1×10^{-4} m^2), and they are separated by a distance of 1 mm (1×10^{-3} m). The permittivity of free space (ε_0) is 8.9×10^{-12} C/Vm. So, you can calculate the capacitance:

$$C = \frac{\varepsilon_0 A}{d}$$

$$C = \frac{(8.9 \times 10^{-12} \text{ C/V} \cdot \text{m})(1 \times 10^{-4} \text{ m}^2)}{(1 \times 10^{-3} \text{ m})}$$

$$C = 8.9 \times 10^{-13} \text{ F}$$

345. **(C)** When wires are cut through a magnetic field, the maximum flux occurs in those that are perpendicular to the direction of the field, and zero will occur in those that are parallel to the field. All others will have intermediate components based on their angle (the greater the angle, the greater the flux). In this diagram, the order of wires from least to greatest is $C < E < D < A < B$.

346. **(D)** The charge is 0.1 C and the distance is 3 m. You can calculate the voltage:

$$V = \frac{kQ}{r^2}$$

$$V = \frac{(9\times10^9\ \text{N}\cdot\text{m}^2/\text{C}^2)(0.1\ \text{C})}{(3\ \text{m})^2}$$

$$V = 1.0\times10^8\ \text{V}$$

347. **(E)** You know the charge is 0.2 C and the force is 100 N. You can calculate the field strength:

$$F = qE$$

$$E = \frac{F}{q}$$

$$E = \frac{(100\ \text{N})}{(0.2\ \text{C})}$$

$$E = 500\ \text{N/C}$$

348. **(E)** The negatively charged object induces a charge in the neutral object. The negatively charged object repels negative charges in the neutral object to the opposite side. The remaining positive charges attract the negatively charged object.

349. **(C)** When placed in a uniform electric field directed to the right, only the proton will move to the right. The neutron and hydrogen atom will be unaffected, while the electron and antiproton will move to the left.

350. **(D)** The potential energy of the 0.05 C charge in the field is 100 J. You can calculate the voltage:

$$PE = qV$$

$$V = \frac{PE}{q}$$

$$V = \frac{(100\ \text{J})}{(0.05\ \text{C})}$$

$$V = 2{,}000\ \text{V}$$

351. **(B)** The electric field is 900 N/C and the distance in the field from the point charge is 10 m. You can calculate the magnitude of the charge:

$$E = \frac{kQ}{r^2}$$

$$kQ = Er^2$$

$$Q = \frac{Er^2}{k}$$

$$Q = \frac{(900\ \text{N/C})(10\ \text{m})^2}{(9\times10^9\ \text{N}\cdot\text{m}^2/\text{C}^2)}$$

$$Q = 1\times10^{-5}\ \text{C} = 10\ \mu\text{C}$$

352. (B) The voltage is 120 V and the strength of the field is 1.2×10^6 V/m. You can calculate the distance (d) of separation between the plates:

$$E = \frac{V}{d}$$

$$d = \frac{V}{E}$$

$$d = \frac{(120 \text{ V})}{(1.2 \times 10^6 \text{ V/m})}$$

$$d = 1 \times 10^{-4} \text{ m} = 0.1 \text{ mm}$$

353. (B) The capacitor is 100 μF (1×10^{-4} F) and the voltage is 1.5 V. You can calculate the charge on the plates of the capacitor:

$$Q = CV$$

$$Q = (1 \times 10^{-4} \text{ F})(1.5 \text{ V})$$

$$Q = 1.5 \times 10^{-4} \text{ C}$$

354. (B) You must first calculate the weight of the proton. Next, use the weight of the proton as the force in the electric field, and the charge on the proton to calculate the strength of the field necessary to equal gravity. The direction of the field must be upward.

$$W = mg$$

$$W = (1.7 \times 10^{-27} \text{ kg})(10 \text{ m/s}^2)$$

$$W = 1.7 \times 10^{-26} \text{ N}$$

$$F = W$$

$$F = qE$$

$$E = \frac{F}{q}$$

$$E = \frac{(1.7 \times 10^{-26} \text{ N})}{(1.6 \times 10^{-19} \text{ C})}$$

$$E = 1 \times 10^{7} \text{ N/C}$$

355. (C) The capacitance is 1 μF (1×10^{-6} F), the distance is 0.2 mm (2×10^{-4} m), and $\varepsilon_0 = 8.9 \times 10^{-12}$ C/Vm.

$$C = \frac{\varepsilon_0 A}{d}$$

$$A = \frac{dC}{\varepsilon_0}$$

$$A = \frac{(2 \times 10^{-4} \text{ m})(1 \times 10^{-6} \text{ C/V})}{(8.9 \times 10^{-12} \text{ C/V} \cdot \text{m})}$$

$$A = 22 \text{ m}^2$$

356. (D) The two electric fields will deflect any charged particles. Of the particles listed, only neutrons are uncharged and will not be affected by the electric fields. Therefore, only neutrons will hit the target.

357. (E) The electric field (10 N/C to the right) exerts a force on the charged object ($q = -0.5$ C, $m = 1.00 \times 10^{-3}$ kg). The negatively charged object will accelerate to the left. You can calculate the force on the object and then use Newton's second law to find the acceleration:

$$F = qE$$
$$F = (-0.5\text{ C})(10\text{ N/C})$$
$$F = -5\text{ N}$$
$$W = Fd$$
$$F = ma$$
$$a = \frac{F}{m}$$
$$a = \frac{(-5\text{ N})}{(1.00 \times 10^{-3}\text{ kg})}$$
$$a = -5{,}000\text{ m/s}^2$$

The negative sign indicates the direction of motion is opposite the direction of the electric field (to the left).

358. (C) The force on the electron is (1×10^6 N) opposite the direction of the field. The charge of an electron is (-1.7×10^{-19} C). You can calculate the magnitude of the electric field:

$$F = qE$$
$$E = \frac{F}{q}$$
$$E = \frac{(1 \times 10^6\text{ N})}{(-1.7 \times 10^{-19}\text{ C})}$$
$$E = -5.9 \times 10^{24}\text{ N/C}$$

Ignore the sign of the field; it only indicates direction.

359. (a) $E = \dfrac{V}{d}$

$$E = \frac{(1.5\text{ V})}{(1\text{ m})}$$
$$E = 1.5\text{ V/m}$$

(b) $F = qE$

$F = (1.60 \times 10^{-19}\ \text{C})(1.5\ \text{N/C})$

$F = 2.4 \times 10^{-19}\ \text{N}$

(c) $F = ma$

$$a = \frac{F}{m}$$

$$a = \frac{(2.4 \times 10^{-19}\ \text{N})}{(1.7 \times 10^{-27}\ \text{kg})}$$

$a = 1.4 \times 10^{8}\ \text{m/s}^2$

(d) $v_f = v_i + at$

$v_f = (0\ \text{m/s}) + (1.4 \times 10^{8}\ \text{m/s}^2)(1 \times 10^{-6}\ \text{s})$

$v_f = 140\ \text{m/s}$

(e) $d = v_i t + \frac{1}{2}at^2$

$d = 0 + \frac{1}{2}(1.4 \times 10^{8}\ \text{m/s}^2)(1.0 \times 10^{-6}\ \text{s})^2$

$d = 7.0 \times 10^{-5}\ \text{m or } 70\ \mu\text{m}$

360. (a) $E = \frac{kQ}{r^2}$

$$E_A = \frac{kQ}{r_A^2} = \frac{(9 \times 10^{9}\ \text{N} \cdot \text{m}^2/\text{C}^2)(1.7 \times 10^{-19}\ \text{C})}{(2 \times 10^{-9}\ \text{m})^2} = 3.8 \times 10^{8}\ \text{N/C}$$

$$E_B = \frac{kQ}{r_B^2} = \frac{(9 \times 10^{9}\ \text{N} \cdot \text{m}^2/\text{C}^2)(1.7 \times 10^{-19}\ \text{C})}{(4 \times 10^{-9}\ \text{m})^2} = 9.6 \times 10^{7}\ \text{N/C}$$

$$E_C = \frac{kQ}{r_C^2} = \frac{(9 \times 10^{9}\ \text{N} \cdot \text{m}^2/\text{C}^2)(1.7 \times 10^{-19}\ \text{C})}{(1.6 \times 10^{-8}\ \text{m})^2} = 6.0 \times 10^{6}\ \text{N/C}$$

The greatest field is experienced by electron A.

(b) $V = \frac{kQ}{r}$

$$V_A = \frac{kQ}{r_A} = \frac{(9\times10^9\ \text{N}\cdot\text{m}^2/\text{C}^2)(1.7\times10^{-19}\ \text{C})}{(2\times10^{-9}\ \text{m})} = 0.77\ \text{V}$$

$$V_B = \frac{kQ}{r_B} = \frac{(9\times10^9\ \text{N}\cdot\text{m}^2/\text{C}^2)(1.7\times10^{-19}\ \text{C})}{(4\times10^{-9}\ \text{m})} = 0.38\ \text{V}$$

$$V_C = \frac{kQ}{r_C} = \frac{(9\times10^9\ \text{N}\cdot\text{m}^2/\text{C}^2)(1.7\times10^{-19}\ \text{C})}{(1.6\times10^{-8}\ \text{m})} = 0.10\ \text{V}$$

The lowest voltage is experienced by electron C.

(c) $F = \frac{kQ_1Q_2}{r^2}$

$$F_A = \frac{kQ_AQ_2}{r_A^2} = \frac{(9\times10^9\ \text{N}\cdot\text{m}^2/\text{C}^2)(-1.7\times10^{-19}\ \text{C})(1.7\times10^{-19}\ \text{C})}{(2\times10^{-9}\ \text{m})^2} = -6.5\times10^{-11}\ \text{N}$$

$$F_B = \frac{kQ_BQ_2}{r_B^2} = \frac{(9\times10^9\ \text{N}\cdot\text{m}^2/\text{C}^2)(-1.7\times10^{-19}\ \text{C})(1.7\times10^{-19}\ \text{C})}{(4\times10^{-9}\ \text{m})^2} = -1.6\times10^{-11}\ \text{N}$$

$$F_C = \frac{kQ_CQ_2}{r_C^2} = \frac{(9\times10^9\ \text{N}\cdot\text{m}^2/\text{C}^2)(-1.7\times10^{-19}\ \text{C})(1.7\times10^{-19}\ \text{C})}{(1.6\times10^{-8}\ \text{m})^2} = -1.0\times10^{-12}\ \text{N}$$

All the electrons experience attractive forces toward the protons. The greatest force is experienced by electron A.

Chapter 13: Circuits

361. (C) The electric charge is 240 C and the time interval is 10 s. Here is the current:

$$I = \frac{\Delta Q}{\Delta t}$$

$$I = \frac{(240\ \text{C})}{(10\ \text{s})}$$

$$I = 24\ \text{A}$$

362. (A) The voltage is 9.0 V and the resistance is 10 Ω. Use Ohm's law to find the current:

$$I = \frac{V}{R}$$

$$I = \frac{(9\ \text{V})}{(10\ \Omega)}$$

$$I = 0.9\ \text{A}$$

363. (E) The voltage is 9 V and the current is 3 A. You can find the power dissipated by the resistor:

$$P = IV$$
$$P = (3\text{ A})(9\text{ V})$$
$$P = 27\text{ J}$$

364. (C) To find the equivalent resistance of resistors in series, add all the resistances together:

$$R_{eq} = R_1 + R_2 + R_3$$
$$R_{eq} = 20\ \Omega + 150\ \Omega + 500\ \Omega$$
$$R_{eq} = 670\ \Omega$$

365. (B) First find the equivalent resistance of the resistors by adding the resistances. Then, use Ohm's law to find the current:

$$R_{eq} = R_1 + R_2 + R_3$$
$$R_{eq} = 3\ \Omega + 1\ \Omega + 2\ \Omega$$
$$R_{eq} = 6\ \Omega$$

$$I = \frac{V}{R_{eq}}$$
$$I = \frac{(12\text{ V})}{(6\ \Omega)}$$
$$I = 2\text{ A}$$

366. (E) In a series circuit, the current is equal in all parts of the circuit. However, the voltage drops across each resistor follow Ohm's law. Since R_1 has the greatest resistance, it has the greatest voltage drop across it.

367. (D) The total power dissipated by the resistors is the product of the total voltage drop (12 V) and the current (2 A):

$$P = IV$$
$$P = (2\text{ A})(12\text{ V})$$
$$P = 24\text{ W}$$

368. (C) Once you have found the equivalent resistance and the total current, use Ohm's law to calculate the voltage drop across the third resistor (2 Ω):

$$R_{eq} = R_1 + R_2 + R_3$$
$$R_{eq} = 3\ \Omega + 1\ \Omega + 2\ \Omega$$
$$R_{eq} = 6\ \Omega$$

$$I = \frac{V}{R_{eq}}$$

$$I = \frac{(12\ \text{V})}{(6\ \Omega)}$$

$$I = 2\ \text{A}$$

$$\Delta V_3 = IR_3$$

$$\Delta V_3 = (2\ \text{A})(2\ \Omega)$$

$$\Delta V_3 = 4\ \text{V}$$

369. (C) In all parts of a series circuit with different resistors, the resistances are different, the voltage drops are different, and the current is the same.

370. (D) When one light in a string of lights goes out and then they all go out, it means the lights are wired in series.

371. (D) In a parallel circuit, the current gets divided among the branches. The amount of current flow is inversely proportional to the resistance.

372. (D) The equivalent resistance of a parallel circuit is calculated as follows:

$$\frac{1}{R_{eq}} = \frac{1}{R_1} + \frac{1}{R_2} + \frac{1}{R_3}$$

$$\frac{1}{R_{eq}} = \frac{1}{(8\ \Omega)} + \frac{1}{(6\ \Omega)} + \frac{1}{(10\ \Omega)}$$

$$\frac{1}{R_{eq}} = 0.4\ \Omega^{-1}$$

$$R_{eq} = 2.6\ \Omega$$

373. (B) Once you have calculated the equivalent resistance, use Ohm's law to find out the total current:

$$\frac{1}{R_{eq}} = \frac{1}{R_1} + \frac{1}{R_2} + \frac{1}{R_3}$$

$$\frac{1}{R_{eq}} = \frac{1}{(8\ \Omega)} + \frac{1}{(6\ \Omega)} + \frac{1}{(10\ \Omega)}$$

$$\frac{1}{R_{eq}} = 0.4\ \Omega^{-1}$$

$$R_{eq} = 2.6\ \Omega$$

$$I = \frac{\Delta V}{R_{eq}}$$

$$I = \frac{(12\text{ V})}{(2.6\ \Omega)}$$

$$I = 4.6\text{ A}$$

374. **(B)** First find the current flowing through the third resistor by using Ohm's law and then calculate the power:

$$I_3 = \frac{\Delta V}{R_3}$$

$$I_3 = \frac{(12\text{ V})}{(10\ \Omega)}$$

$$I_3 = 1.2\text{ A}$$

$$P = IV$$

$$P = (1.2\text{ A})(12\text{ V})$$

$$P = 14\text{ W}$$

375. **(C)** This is a resistor–capacitor (*RC*) circuit. First, calculate the percent charge on the capacitor. Next, determine the time constant of the circuit, which is the product *RC*. Compare the percent charge to the time constant. A capacitor charges to 63% in one time constant and 87% in two time constants.

$$\%C_{total} = \frac{C}{C_{total}} \times 100$$

$$\%C_{total} = \frac{(126\ \mu\text{F})}{(200\ \mu\text{F})} \times 100$$

$$\%C_{total} = 63\%$$

This amount of charge occurs in one time constant (*RC*):

$$\tau = RC$$

$$\tau = (1.00 \times 10^5\ \Omega)(2.00 \times 10^{-4}\text{ F})$$

$$\tau = 20\text{ s}$$

376. **(C)** You calculate the equivalent capacitance of capacitors in series in a manner similar to calculating equivalent resistance for resistors in parallel:

$$\frac{1}{C_{eq}} = \frac{1}{C_1} + \frac{1}{C_2} + \frac{1}{C_3}$$

$$\frac{1}{C_{eq}} = \frac{1}{(5\ \mu\text{F})} + \frac{1}{(4\ \mu\text{F})} + \frac{1}{(2\ \mu\text{F})}$$

$$\frac{1}{C_{eq}} = 0.95\ \mu\text{F}^{-1}$$

$$C_{eq} = 1.05\ \mu\text{F}$$

377. **(E)** You calculate the equivalent capacitance of capacitors in parallel in a manner similar to calculating equivalent resistance for resistors in series:

$$C_{eq} = C_1 + C_2 + C_3$$

$$C_{eq} = 5\ \mu\text{F} + 4\ \mu\text{F} + 2\ \mu\text{F}$$

$$C_{eq} = 11\ \mu\text{F}$$

378. **(A)** The voltage is 9 V and the capacitance is 10 μF. Calculate the charge on the capacitor as follows:

$$Q = CV$$

$$Q = (1.0 \times 10^{-5}\ \text{F})(9.0\ \text{V})$$

$$Q = 9.0 \times 10^{-5}\ \text{C}$$

379. **(C)** First, calculate the current from the charge and time. Then, find the voltage from Ohm's law:

$$I = \frac{Q}{t}$$

$$I = \frac{(40\ \text{C})}{(80\ \text{s})}$$

$$I = 0.5\ \text{A}$$

$$V = IR$$

$$V = (0.5\ \text{A})(10\ \Omega)$$

$$V = 5.0\ \text{V}$$

380. **(E)** Use Ohm's law to find the resistance:

$$V = IR$$

$$R = \frac{V}{I}$$

$$R = \frac{(100\ \text{V})}{(2\ \text{A})}$$

$$R = 50\ \Omega$$

381. (C) Combine Ohm's law and the power equation and then solve for the resistance:

$$P = IV$$
$$V = IR$$
$$P = I(IR)$$
$$P = I^2R$$
$$R = \frac{P}{I^2}$$
$$R = \frac{(1\times10^5 \text{ W})}{(5 \text{ A})^2}$$
$$R = 4\times10^3 \ \Omega = 4 \text{ k}\Omega$$

382. (A) According to Kirchoff's laws, the sum of all the voltages around a closed loop is zero. The battery provides a 9.0 V, while the voltage decreases across each resistor [(–2.7 V) + (–4.2 V) + (–2.1 V) = –9.0 V].

383. (D) The capacitor is 10.0 μF (1 × 10^{-5} F). The capacitor discharges to 37% remaining charge in time constant, τ. You can calculate the resistance from the time constant:

$$\tau = RC$$
$$R = \frac{\tau}{C}$$
$$R = \frac{50 \text{ s}}{1.0\times10^{-5} \text{ C}}$$
$$R = 5.0\times10^6 \ \Omega \text{ or } 5 \text{ M}\Omega$$

384. (E) You must first calculate the equivalent resistance of the resistors in series. Then, use Ohm's law to find the value of the battery voltage:

$$R_{eq} = R_1 + R_2 + R_3$$
$$R_{eq} = 20 \ \Omega + 10 \ \Omega + 30 \ \Omega$$
$$R_{eq} = 60 \ \Omega$$

$$V = IR_{eq}$$
$$V = (2 \text{ A})(60 \ \Omega)$$
$$V = 120 \text{ V}$$

385. (D) Use Ohm's law to calculate the voltage drop across the third resistor (30 Ω) when the current is 5 A:

$$\Delta V_3 = IR_3$$
$$\Delta V_3 = (5 \text{ A})(30 \ \Omega)$$
$$\Delta V_3 = 150 \text{ V}$$

386. **(A)** According to Kirchoff's laws, the sum of all the voltages around a closed loop is zero.

387. **(B)** First, you must calculate the equivalent resistance of the parallel branch of the circuit ($R_{eq||}$). Then, use that to calculate the equivalent resistance of the series circuits:

$$\frac{1}{R_{eq||}} = \frac{1}{R_2} + \frac{1}{R_3}$$

$$\frac{1}{R_{eq||}} = \frac{1}{(10\ \Omega)} + \frac{1}{(20\ \Omega)}$$

$$\frac{1}{R_{eq||}} = 0.15\ \Omega^{-1}$$

$$R_{eq||} = 6.7\ \Omega$$

$$R_{eq} = R_1 + R_{eq||} + R_4$$

$$R_{eq} = 10\ \Omega + 6.7\ \Omega + 45\ \Omega$$

$$R_{eq} = 61.7\ \Omega$$

388. **(B)** Once you have calculated the equivalent resistance of the resistors R_2, R_3, and the parallel branch ($R_{eq||}$), use Ohm's law to calculate the current moving through the circuit. That current must be equal between the first resistor, the parallel branch of resistors, and the final resistor. So, the current leaving the parallel branch must be the same as that which entered it (Kirchoff's law):

$$\frac{1}{R_{eq||}} = \frac{1}{R_2} + \frac{1}{R_3}$$

$$\frac{1}{R_{eq||}} = \frac{1}{(10\ \Omega)} + \frac{1}{(20\ \Omega)}$$

$$\frac{1}{R_{eq||}} = 0.15\ \Omega^{-1}$$

$$R_{eq||} = 6.7\ \Omega$$

$$R_{eq} = R_1 + R_{eq||} + R_4$$

$$R_{eq} = 10\ \Omega + 6.7\ \Omega + 45\ \Omega$$

$$R_{eq} = 61.7\ \Omega$$

$$I = \frac{\Delta V}{R_{eq}}$$

$$I = \frac{(120\ \text{V})}{(61.7\ \Omega)}$$

$$I = 1.9\ \text{A}$$

389. (a) You must calculate the equivalent resistance of the parallel branch of the circuit ($R_{eq||}$), then use that to calculate the equivalent resistance of the series circuits:

$$\frac{1}{R_{eq||}}=\frac{1}{R_2}+\frac{1}{R_3}+\frac{1}{R_4}$$

$$\frac{1}{R_{eq||}}=\frac{1}{(5.0\ \Omega)}+\frac{1}{(2.5\ \Omega)}+\frac{1}{(20\ \Omega)}$$

$$\frac{1}{R_{eq||}}=0.65\ \Omega^{-1}$$

$$R_{eq||}=1.54\ \Omega$$

$$R_{eq}=R_1+R_{eq||}$$

$$R_{eq}=10\ \Omega+1.54\ \Omega$$

$$R_{eq}=11.5\ \Omega$$

(b) Use Ohm's law, the voltage, and the equivalent resistance to calculate the current flowing through the circuit:

$$I=\frac{\Delta V}{R_{eq}}$$

$$I=\frac{(120\ \text{V})}{(11.5\ \Omega)}$$

$$I=10.4\ \text{A}$$

(c) $\Delta V = IR$

$$\Delta V_1 = IR_1 = (10.4\ \text{A})(10\ \Omega) = 104\ \text{V}$$

$$\Delta V_2 = \Delta V_3 = \Delta V_4 = IR_{eq||} = (10.4\ \text{A})(1.54\ \Omega) = 16\ \text{V}$$

(d) $I=\frac{\Delta V}{R}$

$$I_1=\frac{\Delta V_1}{R_1}=\frac{(104\ \text{V})}{(10\ \Omega)}=10.4\ \text{A}$$

$$I_2=\frac{\Delta V_2}{R_2}=\frac{(16\ \text{V})}{(5\ \Omega)}=3.2\ \text{A}$$

$$I_3=\frac{\Delta V_3}{R_3}=\frac{(16\ \text{V})}{(2.5\ \Omega)}=6.4\ \text{A}$$

$$I_4=\frac{\Delta V_4}{R_4}=\frac{(16\ \text{V})}{(20\ \Omega)}=0.8\ \text{A}$$

390. (a) The equivalent resistance increases linearly as you increase the number of resistors in series. In contrast, the equivalent resistance decreases exponentially as you increase the number of resistors in parallel.

(b) The total current decreases exponentially as you increase the number of resistors in series. In contrast, the current increases linearly as you increase the number of resistors in parallel.

(c) The total voltage put out by the battery remains the same as you add resistors either in series or in parallel.

(d) In the series circuit, the power decreases exponentially with increasing number of resistors. The current behaves the same way. In contrast, in the parallel circuit, the power increases linearly with the number of resistors. The current behaves the same way.

(e) As the number of light bulbs in the string increases in the series string of lights, both the current and the power of the circuit decrease. The lights appear dimmer as the number of bulbs increases. For a parallel string of lights, both the current and the power of the circuit increase as the number of light bulbs in the string increases. The lights appear brighter as the number of bulbs increase (although this happens within limits).

Chapter 14: Magnetism

391. (B) The charge of an electron is 1.6×10^{-19} C, the electron's velocity is 1×10^{7} m/s, and the strength of the magnetic field is 1.0 T. You can calculate the force as follows:

$$F = qvB$$

$$F = (1\times10^{-19}\ \text{C})(1\times10^{7}\ \text{m/s})(1.0\ \text{T})$$

$$F = 1.6\times10^{-12}\ \text{N}$$

392. (A) For a magnetic field to exert a force on a moving electron, the direction of the electron's movement must have components that are perpendicular to the direction of the magnetic field. Of the angles listed, the electron moving at 0° has no component that is perpendicular to the direction of magnetic field. Therefore, the magnetic field exerts no force upon it.

393. (C) The proton moves at an angle of 30° and a speed of 5×10^{6} m/s relative to the magnetic field. The field exerts a force of 4.8×10^{-13} N on the proton. You can calculate the strength of the magnetic field as follows:

$$F = qvB\sin\theta$$

$$B = \frac{F}{qv\sin\theta}$$

$$B = \frac{(4.8\times10^{-13}\ \text{N})}{(1.6\times10^{-19}\ \text{C})(5\times10^{6}\ \text{m/s})\sin(30°)}$$

$$B = 1.2\ \text{T}$$

394. (C) The magnetic field is directed into the paper and the velocity vector of the charge is to the right. If you apply right-hand rule no. 1, you will see that the magnetic field exerts an upward force on the charge.

395. (A) First find the equivalent resistance of the resistors by adding the resistances. Then, use Ohm's law to find the current.

$$F = qvB$$

$$F = (1\times10^{-6}\ \text{C})(5\times10^{3}\ \text{m/s})(10\ \text{T})$$

$$F = 5\times10^{-2}\ \text{N or } 0.05\ \text{N}$$

396. (D) If the charge shown in the figure was negative instead of positive, then the magnetic field would exert a force on the charge in the opposite direction (i.e., downward).

397. (B) By applying right-hand rule no. 1 to the charge at various positions in the circle, you can see that the magnetic field exerts a centripetal force on the charge.

398. (A) The magnetic field is directed down and the current vector is out of the paper. If you apply right-hand rule no. 1, you will see that the magnetic field exerts a force to the right on the wire.

399. (C) If the current is 10 A, the magnetic field strength is 10 T, and the length of wire inside the magnetic field is 1 cm or 0.01 m, then you can calculate the magnitude of the force:

$$F = ILB$$

$$F = (10\ \text{A})(0.01\ \text{m})(10\ \text{T})$$

$$F = 1.0\ \text{N}$$

400. (D) By applying right-hand rule no. 1 to the situation, the magnetic field will exert a force on the water to the right. Newton's third law indicates that the reaction force (thrust) is the opposite of the magnetic force. So, the thrust will be to the left.

401. (E) By applying right-hand rule no. 2 to the diagram, you see that the magnetic field comes out of the paper to the left of the wire and goes into the paper to the right of the wire.

402. (E) The point that is closest to the center of the wire (i.e., the shortest radius) will experience the strongest magnetic field. In the figure, Point E is closest to the wire's center.

403. (C) The current in the wire is 5 A, the radius from the center of the wire is 5 mm (5×10^{-3} m), and the permeability of free space is $4\pi \times 10^{-7}\ \text{T}\cdot\text{m/A}$. The strength of the magnetic field can be calculated:

$$B = \frac{\mu_0 I}{2\pi r}$$

$$B = \frac{(4\pi\times10^{-7}\ \text{T}\cdot\text{m/A})(5\ \text{A})}{2\pi(5\times10^{-3}\ \text{m})}$$

$$B = 2\times10^{-4}\ \text{T}$$

404. (B) The charge is 1 μC (1×10^{-6} C) at a distance of 5 mm (5×10^{-3} m) from the center of the wire. It is moving at 5 m/s in the same direction as the current (5 A). You can calculate the magnitude of the force:

$$F = qvB$$

$$B = \frac{\mu_0 I}{2\pi r}$$

$$F = qv\left(\frac{\mu_0 I}{2\pi r}\right)$$

$$F = (1\times10^{-6}\ \text{C})(5\ \text{m/s})\left[\frac{(4\pi\times10^{-7}\ \text{T}\cdot\text{m/A})(5\ \text{A})}{2\pi(5\times10^{-3}\ \text{m})}\right]$$

$$F = 1\times10^{-9}\ \text{N}$$

405. (C) If the solenoid has 100 coils/m and the wire carries 10 A of current, then you can calculate the strength of the magnetic field:

$$B = \mu_0 nI$$

$$B = (4\pi\times10^{-7}\ \text{T}\cdot\text{m/A})(100\ \text{m}^{-1})(10\ \text{A})$$

$$B = 1.3\times10^{-3}\ \text{T}$$

406. (D) The conducting rod ($L = 2.0$ m) moves at a velocity of 10 m/s. The magnetic field is 2T. You can calculate the magnitude of the electromagnetic field (EMF; ε):

$$\varepsilon = vBL$$

$$\varepsilon = (10\ \text{m/s})(2\ \text{T})(2.0\ \text{m})$$

$$\varepsilon = 40\ \text{V}$$

407. (A) The magnetic field is 0.5 T, and the area that it passes through is 6.3 cm^2 ($6.3 \times 10^{-4}\ \text{m}^2$).

$$\phi = BA$$

$$\phi = (0.5\ \text{T})(6.3\times10^{-4}\ \text{m}^2)$$

$$\phi = 3.2\times10^{-4}\ \text{Wb}$$

408. (D) The number of coils (N) on one side of a transformer was 100. The magnetic flux changed at a rate of 10 Wb/s. The direction of the change was not specified, so you can only calculate the magnitude of the change using Faraday's law:

$$\varepsilon = -N\frac{\Delta\phi}{\Delta t}$$

$$\varepsilon = -(100)(10\ \text{Wb/s})$$

$$\varepsilon = -1{,}000\ \text{V}$$

Because the direction of the changing magnetic flux was not specified, the magnitude of $\varepsilon = 1{,}000$ V.

409. (D) The magnetic field is 1×10^{-4} T. The point is 5 cm (5×10^{-2} m) away from the wire. You can calculate the current in the wire:

$$B = \frac{\mu_0 I}{2\pi r}$$

$$\mu_0 I = B 2\pi r$$

$$I = \frac{B 2\pi r}{\mu_0}$$

$$I = \frac{(1 \times 10^{-4}\ \text{T}) 2\pi (5 \times 10^{-2}\ \text{m})}{(4\pi \times 10^{-7}\ \text{T} \cdot \text{m/A})}$$

$$I = 25\ \text{A}$$

410. (C) Apply right-hand rule no. 2 to the wire on the left and you will see the direction of the magnetic field as shown in this figure. Now, the magnetic field of the left wire exerts a force on the right wire. You can determine the direction of this force by applying right-hand rule no. 1 to the left wire. You will find that the left wire exerts a force on the right wire, pushing it to the right as shown in the figure. According to Newton's third law, the right wire's magnetic field will push the left wire to the left. The result is that the two wires will repel each other.

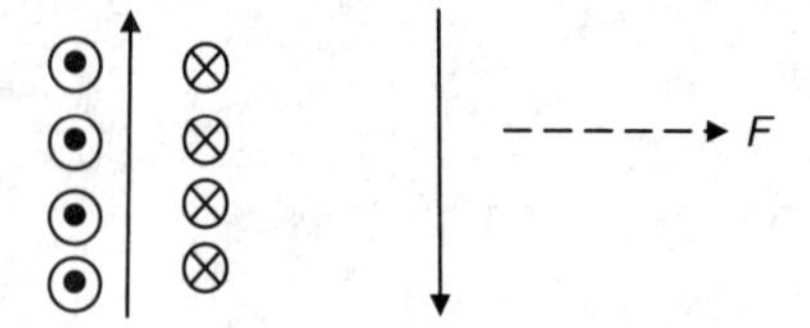

411. (B) The current is 2 A, the length of wire in the field is 10 cm (0.1 m), and the force is 0.1 N. You can calculate the strength of the magnetic field:

$$F = ILB$$

$$B = \frac{F}{IL}$$

$$B = \frac{(0.1\ \text{N})}{(2\ \text{A})(0.1\ \text{m})}$$

$$B = 0.5\ \text{T}$$

412. (A) The strength of the magnetic field is 1.0×10^{10} T and it exerts a force of 1.6×10^{-3} N on the charge. The charge is moving at 1×10^{6} m/s perpendicular to the field. You can calculate the magnitude of the charge:

$$F = qvB$$

$$q = \frac{F}{vB}$$

$$q = \frac{(1.6 \times 10^{-3}\ \text{N})}{(1.0 \times 10^{6}\ \text{m/s})(1 \times 10^{10}\ \text{T})}$$

$$q = 1.6 \times 10^{-19}\ \text{C}$$

413. (A) The magnetic field produced by a wire-wrapped torus is zero everywhere outside the torus.

414. (A) By applying right-hand rule no. 2 to the figure, you can determine that the current is moving up the wire.

415. (D) The strength of the magnetic field is 5×10^{-3} T at 1 mm (1×10^{-3} m) away. You can calculate the magnitude of the current:

$$B = \frac{\mu_0 I}{2\pi r}$$

$$\mu_0 I = B2\pi r$$

$$I = \frac{B2\pi r}{\mu_0}$$

$$I = \frac{(5\times10^{-3}\ \text{T})2\pi(1\times10^{-3}\ \text{m})}{(4\pi\times10^{-7}\ \text{T}\cdot\text{m/A})}$$

$$I = 25\ \text{A}$$

416. (D) The charge is 1 μC (1.0×10^{-6} C). It moves opposite the direction of the current at 1 cm (1×10^{-2} m) away from the center of the wire. The charge moves at a velocity of 5 m/s. You can calculate the magnitude of the force exerted on the charge by the wire's magnetic field:

$$F = qvB$$

$$B = \frac{\mu_0 I}{2\pi r}$$

$$F = qv\left(\frac{\mu_0 I}{2\pi r}\right)$$

$$F = (1\times10^{-6}\ \text{C})(5\ \text{m/s})\left[\frac{(4\pi\times10^{-7}\ \text{T}\cdot\text{m/A})(25\ \text{A})}{2\pi(1\times10^{-2}\ \text{m})}\right]$$

$$F = 2.5\times10^{-9}\ \text{N}$$

You can calculate the direction of the force exerted by the magnetic field with right-hand rule no. 1 and determine that it is away from the center of the wire.

417. (E) You can use right-hand rule no. 1 to determine the direction of the force exerted upon the charge. Remember, this is a negative charge, so the direction of the force will be opposite the palm of your hand. In this case the direction is 315°.

418. (C) The –1 μC (-1×10^{-6} C) charge moves at an angle of 45° and a velocity of 10 m/s. The strength of the magnetic field is 10 T. You can calculate the magnitude of the force exerted on the charge (ignore the sign):

$$F = qvB\sin\theta$$

$$F = (1\times10^{-6}\ \text{C})(10\ \text{m/s})(10\ \text{T})\sin 45^\circ$$

$$F = (1\times10^{-6}\ \text{C})(10\ \text{m/s})(10\ \text{T})(0.71)$$

$$F = 7\times10^{-5}\ \text{N}$$

419. (a) Here's how to calculate the magnitude of the induced EMF (ε):

$$\varepsilon = vBL$$

$$\varepsilon = (5\text{ m/s})(2\text{ T})(2.0\text{ m})$$

$$\varepsilon = 20\text{ V}$$

(b) Use Ohm's law to calculate the current flowing through the circuit:

$$I = \frac{\varepsilon}{R}$$

$$I = \frac{(20\text{ V})}{(100\ \Omega)} = 0.2\text{ A}$$

(c) The power dissipated by the light bulb can be calculated as follows:

$$P = I\varepsilon$$

$$P = (0.2\text{ A})(20\text{ V})$$

$$P = 4\text{ W}$$

(d) The energy used by the light bulb in 60 s can be calculated as follows:

$$E = Pt$$

$$E = (4\text{ W})(60\text{ s})$$

$$E = 240\text{ J}$$

420. (a) The time constant of the circuit is the time when 63% of the total current has built up ($0.63 \times 1.2\text{ A} = 0.76\text{ A}$). Now, read the time when the current is 0.76 A to get the time constant. The value is 5 ms.

(b) Calculate the value of the inductance, L, from the time constant (5 ms or 5×10^{-3} s) and the value of the resistance (10 Ω):

$$\tau = \frac{L}{R}$$

$$L = \tau R$$

$$L = (5 \times 10^{-3}\text{ s})(10\ \Omega)$$

$$L = 5 \times 10^{-2}\text{ H or } 50\text{ mH}$$

(c) Read the current from the graph at 2 τ (10 ms). The current is 1.04 A.

(d) Calculate the voltage drop across the resistor at 10 ms using Ohm's law where the current is 1.04 A and the resistance is 10 Ω:

$$I = \frac{\Delta V}{R}$$

$$\Delta V = IR$$

$$\Delta V = (1.04\text{ A})(10\ \Omega)$$

$$\Delta V = 10.4\text{ V}$$

Chapter 15: Waves

421. (C) Of the waves listed, only sound is an example of a longitudinal wave in which the particles vibrate in the direction of the wave.

422. (E) The amplitude of a wave is the distance between the crest and the horizontal axis, or the trough and the horizontal axis.

423. (B) The period of a wave whose frequency is 100 Hz can be calculated as follows:

$$T = \frac{1}{f}$$

$$T = \frac{1}{(100 \text{ Hz})}$$

$$T = 0.01 \text{ s}$$

424. (D) Here's how to calculate the velocity of a wave with a frequency of 100 Hz and a wavelength of 1.0 m:

$$v = f\lambda$$

$$v = (100 \text{ Hz})(1.0 \text{ m})$$

$$v = 100 \text{ m/s}$$

425. (D) A sound wave travels at 343 m/s. The wavelength of the sound wave is 17.2 cm (1.72×10^{-2} m). You can calculate the frequency:

$$v = f\lambda$$

$$f = \frac{v}{\lambda}$$

$$f = \frac{(343 \text{ m/s})}{(1.72 \times 10^{-2} \text{ m})}$$

$$f = 1.99 \times 10^4 \text{ Hz or } 20 \text{ kHz}$$

426. (C) The sound wave broadcast through the headphones is 180° out of phase with the one from the jackhammer. The two waves interfere with each other and cancel out. The result is that the operator does not hear the jackhammer noise. This is an example of destructive interference.

427. (D) The length of the open tube is 0.5 m. The speed of sound at 20°C is 343 m/s. You can calculate the fundamental frequency of the tube ($n = 1$):

$$f_n = \frac{nv}{2L}$$

$$n = 1$$

$$f_1 = \frac{(1)(343 \text{ m/s})}{2(0.5 \text{ m})}$$

$$f_1 = 343 \text{ Hz}$$

428. (B) The wavelength is 475 nm or 4.75×10^{-7} m. The spacing in the diffraction grating is 0.5 mm or 5×10^{-4} m. The screen is 2 m away. Here's how to calculate the distance from the center spot (x):

$$x = \frac{m\lambda L}{d}$$

$$x = \frac{(2)(4.75\times10^{-7}\ \text{m})(2\ \text{m})}{(5\times10^{-4}\ \text{m})}$$

$$x = 3.8\times10^{-3}\ \text{m or } 0.38\ \text{mm}$$

429. (C) The wavelength of light entering the material is 510 nm, while the wavelength of light within the material is 638 nm. You can calculate the index of refraction of the material:

$$\lambda_n = \frac{\lambda}{n}$$

$$n = \frac{\lambda}{\lambda_n}$$

$$n = \frac{(510\ \text{nm})}{(638\ \text{nm})}$$

$$n = 0.80$$

430. (D) Maxwell's third equation is Faraday's law.

431. (D) The two waves will add together and the resulting wave will have the amplitude 2A/3.

432. (C) The figure depicts the phenomenon of destructive interference.

433. (D) After the waves meet, they will travel on as depicted in the diagram. One wave (+A) will travel to the right, while one wave (–A/3) will travel to the left.

434. (B) The frequency of the tuning fork is 440 Hz. At the temperature specified, the speed of sound is 343 m/s. We are looking for the length, L, of the air column when the first ($n = 1$) resonance sound is heard.

$$f_n = \frac{nv}{4L}$$

$$L = \frac{nv}{4f_n}$$

$$L = \frac{(1)(343\ \text{m/s})}{4(440\ \text{Hz})}$$

$$L = 0.19\ \text{m}$$

435. (C) The tsunami wave's velocity is 720 km/h (200 m/s). The period of the wave is 10 min (600 s). You can calculate the wavelength:

$$v = \frac{\lambda}{T}$$

$$\lambda = vT$$

$$\lambda = (200 \text{ m/s})(600\text{s})$$

$$\lambda = 1.2 \times 10^5 \text{ m or } 120 \text{ km}$$

436. (E) According to Maxwell, only accelerating charged particles will produce an electromagnetic wave (i.e., light). An accelerating proton is such a particle.

437. (B) The wavelength of light entering the gasoline is 590 nm. The index of refraction of gasoline is 1.4. You can calculate the wavelength of light within the gasoline:

$$\lambda_n = \frac{\lambda}{n}$$

$$\lambda_n = \frac{(590 \text{ nm})}{(1.4)}$$

$$\lambda_n = 421 \text{ nm}$$

438. (B) The speed of light, c, in a vacuum is 3.0×10^8 m/s. The index of refraction, n, of the glass is 1.5. You can calculate the speed of the light wave within the glass:

$$n = \frac{c}{v}$$

$$v = \frac{c}{n}$$

$$v = \frac{(3.0 \times 10^8 \text{ m/s})}{(1.5)}$$

$$v = 2.0 \times 10^8 \text{ m/s}$$

439. (B) A monochromatic light wavelength of 500 nm passes through a 5-μm slit and produces an interference pattern on a screen. You can calculate the angle from the slit where you would find the fifth spot ($m = 5$) from the center:

$$d \sin \theta = m\lambda$$

$$\sin \theta = \frac{m\lambda}{d}$$

$$\sin \theta = \frac{(5)(5 \times 10^{-7} \text{ m})}{5 \times 10^{-6} \text{ m}}$$

$$\sin \theta = 0.5$$

$$\theta = 30^\circ$$

440. (B) Of the EMF waves listed, radio waves have the lowest frequency and longest wavelengths.

441. (A) The speed of a radio wave is the speed of light (3×10^8 m/s) and the frequency is 100 MHz (1×10^8 Hz). You can calculate the wavelength of the radio wave:

$$v = f\lambda$$

$$\lambda = \frac{v}{f}$$

$$\lambda = \frac{(3\times10^8 \text{ m/s})}{(1\times10^8 \text{ Hz})}$$

$$\lambda = 3 \text{ m}$$

442. (D) The change in frequency of a wave emitted from a moving object as that object passes you is called the Doppler effect.

443. (A) The pipe length is 3 m and is closed at one end. You can calculate the frequency of the third harmonic ($n = 3$):

$$f_n = \frac{nv}{4L}$$

$$f_3 = \frac{(3)(343 \text{ m/s})}{4(3 \text{ m})}$$

$$f_3 = 86 \text{ Hz}$$

444. (B) The beat frequency is the difference between the two frequencies of the tuning forks (440 Hz – 320 Hz = 120 Hz).

445. (A) The medium above the top surface of the soap film and that below the bottom surface are identical. In this instance, there is only one phase reversal in the waves that are reflected from the top and bottom surfaces of the film. So, constructive interference occurs when the extra distance traveled by the ray through the film is one-half the wavelength of within the film (λ_n). Therefore, with an index of refraction, n, of 1.5 and a wavelength of 600 nm perpendicular to the film, solve for the minimum thickness of the film at which constructive interference occurs:

$$2t = \left(m + \frac{1}{2}\right)\lambda_n$$

$$n = \frac{\lambda}{\lambda_n}$$

$$\lambda_n = \frac{\lambda}{n}$$

$$2t = \left(m + \frac{1}{2}\right)\frac{\lambda}{n}$$

$$2t = \left(m+\frac{1}{2}\right)\lambda_n$$

$$n = \frac{\lambda}{\lambda_n}$$

$$\lambda_n = \frac{\lambda}{n}$$

$$2t = \left(m+\frac{1}{2}\right)\frac{\lambda}{n}$$

$2nt = \left(m+\frac{1}{2}\right)\lambda$, where m = 0, 1, 2. We are interested in $m = 0$.

$$2nt = \left(0+\frac{1}{2}\right)\lambda$$

$$2nt = \frac{\lambda}{2}$$

$$t = \frac{\lambda}{4n}$$

$$t = \frac{(600 \text{ nm})}{4(1.5)}$$

$$t = 100 \text{ nm}$$

446. (C) The two waves will add together and the resulting wave will have the amplitude 3A/2.

447. (E) A wavelength can be measured either from crest to crest or trough to trough.

448. (C) The light travels through the water at a velocity of 2.26×10^8 m/s. You can calculate the index of refraction of the water:

$$n = \frac{c}{v}$$

$$n = \frac{(3.0\times10^8 \text{ m/s})}{(2.3\times10^8 \text{ m/s})}$$

$$n = 1.3$$

449. (a) The relationship between the order number and the distance from the center of the spot is linear (this is shown in all of the graphs). For any given wavelength and slit spacing, the interval between each spot is constant.

(b) As you increase the spacing between the slits, the distance of the spots from the center decreases (i.e., the spots get closer together). This relationship is shown in **Experiment 1**. The slit spacing is probably the most influential factor in determining the separation of the spots.

(c) As you increase the wavelength, the spacing between the spots increases. This is shown in **Experiment 2**.

(d) The data from **Experiment 1** shows that decreasing the slit spacing will increase the spacing between diffraction bands (spots for lasers). For example, a diffraction grating with a small spacing (on the order of 1–100 μm) will give a good separation.

450. As the light ray passes through the oil film, there are two phase reversals, one between air and oil and the other between oil and glass. So, destructive interference occurs (i.e., the blue light gets subtracted out) when the extra distance traveled by the ray through the film is one-half the wavelength of within the film (λ_n). Therefore, with an index of refraction, n, of 1.3 and a wavelength of 475 nm perpendicular to the film, solve for the minimum thickness of the film at which constructive interference occurs:

$$2t = \left(m + \frac{1}{2}\right)\lambda_n$$

$$n = \frac{\lambda}{\lambda_n}$$

$$\lambda_n = \frac{\lambda}{n}$$

$$2t = \left(m + \frac{1}{2}\right)\frac{\lambda}{n}$$

$$2nt = \left(m + \frac{1}{2}\right)\lambda, \text{ where } m = 0, 1, 2. \text{ We are interested in } m = 0$$

$$2nt = \left(0 + \frac{1}{2}\right)\lambda$$

$$2nt = \frac{\lambda}{2}$$

$$t = \frac{\lambda}{4n}$$

$$t = \frac{(475 \text{ nm})}{4(1.3)}$$

$$t = 91.3 \text{ nm}$$

Chapter 16: Optics

451. (D) Red, orange, yellow, green, blue, and violet.

452. (C) With the indices the same, light is not refracted, and the glass is not seen.

453. (A) 50 cm is obtained from $f = R/2$.

454. (D) The focal length is one-half of the radius of curvature or 4 km. The reflected light would have to be focused on the ships at sea 2 km away. Given the ancient Greeks' knowledge of geometry, they could very likely make such mirrors.

455. (B) This type of lens would move the eye's focal point back toward the retina from inside of the eye.

456. (C) This can be done using two lenses of different refractivity such as a convex-convex crown glass lens mated with a concave-concave flint glass lens.

457. (D) Because of the law of reflection, the reflected beam comes off at the same angle as the incident beam.

458. (D) A matte surface would diffuse the laser beam and minimize the likelihood of burning something or causing an injury.

459. (C) This is because the line of sight of the fish bends slightly away from normal when it moves from water to air.

460. (D) It should be lower than that of the fiber so that total reflection can take place.

461. (D) All of the answers are correct.

462. (E) The magnification is found from dividing the object lens's focal length by the eyepiece's focal length, so that the magnification = 980/50 = 19.6.

463. (E) Newton's telescope did not eliminate spherical aberration.

464. (D) The lack of atmospheric aberration contributes to the clarity of the image.

465. (D) It is above much of the atmosphere and has low light pollution, but it only views the Southern Hemisphere. It is also in a desert with low humidity and is electronically connected to far-away astronomers.

466. (E) Shorter wavelengths are refracted more than longer wavelengths.

467. (B) An image in a mirror will appear as far behind the mirror as the object in front of it. Theoretically, there should be an infinity of images with two parallel mirrors going into the distance in the mirror. But, because of the resolution power of the eyes, there will be a point at which the reflected image will be too small to be seen.

468. (B) Using the mirror formula, $1/p + 1/q = 1/f = 2/R$, p is considered to be infinity and thus the star images will appear at the focal length 200 cm.

469. (C) This lens will bring the eye's focal point back to the retina from behind the eye.

470. (E) A cylindrical lens is asymmetric and can be used to correct for asphericity in the eye.

471. **(E)** A one-way mirror works because the observers' room is darkened, the person being observed is brightly lit, and the observer side of the mirror is only partially silvered. All three conditions must be met. Glass does have a large index of refraction.

472. **(C)** Drops of water act like lenses and focus the sun on the leaf. Thus causing it to burn and turn brown.

473. **(B)** A cat's eyes are mirrored.

474. From the geometry, the observer's line of sight when the cup is empty forms the hypotenuse of a right triangle with sides, d (depth of cup), and 6 cm (diameter of the cup). The hypotenuse also forms the angle, α, with the side of the cup. Similarly, when the cup is filled with water, the observer's line of sight is refracted by the water and forms the hypotenuse of another triangle with sides d and 3 cm. Let the refracted angle with the wall of the cup be β. Then, use the Pythagorean theorem, $\sin\alpha = 6/\sqrt{36+d^2}$ and $\sin\beta = 3/\sqrt{9+d^2}$. Then, use the laws of refraction, $\sin\alpha/\sin\beta = 6/\sqrt{36+d^2}/3/\sqrt{9+d^2}$. When this ratio equals 1.333, the refractive index of water, d, is 3.05 cm.

475. Using the lens equation, $1/d_{obj} + 1/d_{im} = 1/f$, f and d_{obj} are known and so d_{im} can be found. They are respectively 30 cm, 40 cm, ∞, -20 cm, and -6.7 cm.

Chapter 17: Atomic and Nuclear Physics

476. **(C)** Divide the critical mass by the density to get the volume. Find the diameter of the sphere from the formula for the volume of a sphere.

477. **(A)** The naturally occurring radioactive elements are: technetium, promethium, polonium, astatine, radon, francium, radium, actinium, thorium, protactinium, uranium, and neptunium. The manmade elements are plutonium, americium, curium, berkelium, californium, einsteinium, fermium, mendelevium, nobelium, lawrencium, rutherfordium, dubnium, seaborgium, bohrium, hassium, meitnerium, darmstadtium, roentgenium, copernicium, ununtrium, ununquadium, ununpentium, ununhexium, ununseptium, and ununoctium.

478. **(B)** They make up a helium nucleus, an electron, and a photon.

479. **(B)** A helium nucleus, an electron, and a photon.

480. **(E)** 15×10^8 J is found from $E = mc^2$.

481. **(D)** 92, 92, 143 in which the atomic number, 92, provides the number of electrons and protons, and the atomic number subtracted from the isotope number yields the number of neutrons.

482. **(D)** The energy in a photon is given by the equation $h \times c/\lambda$, in which h and c are constants. The wavelength, λ, is the only variable. The frequency, while variable, also depends on the wavelength.

483. (C) Half-life is the period of time it takes for a substance undergoing decay to decrease by half. A simple formula for this is:

$$N(t) = N_0\left(\frac{1}{2}\right)^{t/t_{1/2}}$$

N is the quantity of the substance that varies with time, t is the variable time, and $t_{1/2}$ is the half-life of the substance.

484. (A) Alpha particles contain two protons and two neutrons. To reduce the atomic weight from 238 to 206, eight alpha particles must be emitted, which remove 32 neutrons and protons.

485. (B) 2.02 eV is found by using the equation for the energy of a photon, $E = h \times c/\lambda$, and noting significant figures.

486. (D) The slow neutron has no charge and can hit the nucleus to cause fission, while a slow proton has a positive electric charge and is repulsed by the much larger positive charge of the uranium nucleus.

487. (D) Most elements are a mix of isotopes with more or less neutrons than predicted by the atomic number subtracted from the amu. The neutrons account for the fractional amus.

488. (E) The electron and positron are of opposite charge and annihilate each other by emitting two photons to conserve energy.

489. (B) An alpha particle is $_2$helium^4 with no electrons. When $_7$nitrogen^{14} is bombarded, the elementary particles are conserved and usually stable isotopes are produced. In this case, $_1$hydrogen^1 is produced so the remaining neutron and proton must go to produce $_8$oxygen^{17}. $_1$hydrogen^3 is radioactive $_1$tritium^3; $_1$hydrogen^2 is radioactive $_1$deuterium^2, making (A) and (C) wrong. In (D) and (E) the numbers do not add up correctly.

490. (D) The energies are calculated from the energy of a photon equation: $E = h \times c/\lambda$, while 1 eV = 1.6×10^{-19} J/eV accounts for the units.

491. (D) There are 6.022×10^{23} atoms/mol of substance. The mass of 1 mol equals the atomic mass of the element, in this case 235 g or 0.235 kg. One eV is equal to 1.6×10^{-19} J. Then the number of atoms, multiplied by the energy emitted, divided by the atomic mass gives the result when units are accounted for.

492. (B) One eV is equal to 1.6×10^{-19} J/eV. An electron has a mass of 9.11×10^{-31} kg. The kinetic energy of the electron is 1.6×10^{-19} J. From the kinetic energy the velocity can be calculated by kinetic energy = one-half $\times$ $m \times v^2$. Solving for v yields 5.93×10^5 m/s when accounting for the units.

493. (B) A photon is a quantum of energy, which allows the effect.

494. (C) Although it also was related to the other answers, (C) was the most important.

495. (C) Light behaves like a wave or a particle depending on which property is being examined. This is a mathematical construct to allow examination of the phenomenon of light and its interaction with the natural world.

496. (D) Neutrons hitting atoms cause fissions. These can be regulated by neutron absorbers.

497. (D) Most elements can be split, but some like hydrogen and helium cannot because they do not have the proper number of particles to cause alpha-particle emission.

498. (E) In beta decay, the atom's charge and kinetic energy must be conserved. So depending on the type of beta decay, a neutron turns into a proton with emission of an electron. Similarly in beta decay in which a positron is emitted, a proton turns into a neutron to maintain the proper charge. During beta decay a neutrino and an antineutrino are emitted to conserve the system's kinetic energy.

499. We are given the radius of the universe shortly after the Big Bang: converted to centimeters it is 1.5×10^{13} cm. The volume of a sphere is $0.75 \times \pi \times r^3$, where r is the radius of the sphere, which in this case is 1.5×10^{13} cm. So, the volume is $3.53 \times 10^{-30}\text{cm}^3$. We can calculate the average mass of the three particles, $(P + N + e)/3$, or, when converted to grams, $(1.67 \times 10^{-30}\,\text{g} + 1.67 \times 10^{-30}\,\text{g} + 9.11 \times 10^{-33}\,\text{g})/3 = 1.11 \times 10^{-30}$. We were given the mass density of the sphere shortly after the Big Bang to be 10^{15} g/cm^3. Finally, we can divide the mass density by the average particle mass to get the number of particles in the sphere, or $1.41 \times 10^{49}/1.11 \times 10^{-30} = 1.27 \times 10^{79}$.

500. 9.4×10^{19} fissions/s and 2.88×10^{-3} kg/day. A watt is a J/s. So, 3.000×10^6 J/s = X number of fissions, $\times\ 200 \times 10^6$ eV per fission, $\times\ 1.6 \times 10^{-19}$ J/eV yields: 9.4×10^{19} fissions/s. From $E = mc^2$, 3.000×10^9 J/s = X kg $\times\ (3.00 \times 10^8)^2$. Solving for $X = 3.000 \times 10^9$ J/s $\times$ 60 s/m $\times$ 60 m/hr $\times$ 24 hr/day/$(3.00 \times 10^8)^2$ yields 2.88×10^{-3} kg/day.